NOTE:-

DAMAGED TUBES MAY BE REPAIRED BY REPLACING
THE DAMAGED PORTION WITH NEW LENGTH OF
TUBE CONNECTED TO EXISTING TUBE WITH A.G.S. 838.

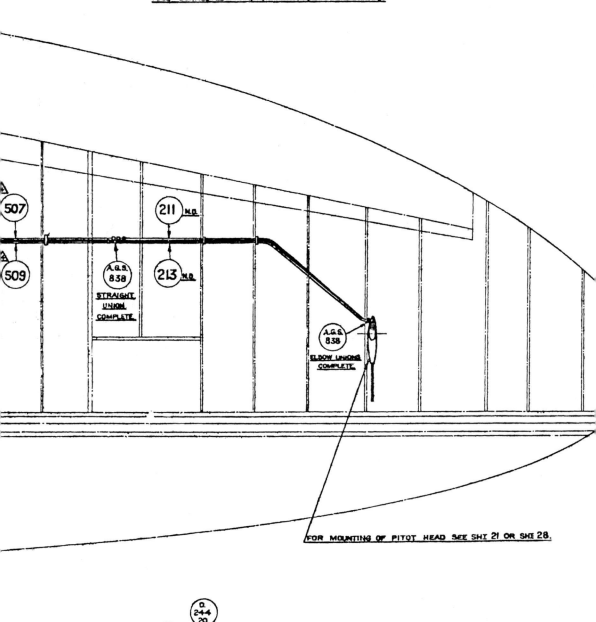

FOR MOUNTING OF PITOT HEAD SEE SHT 21 OR SHT 28.

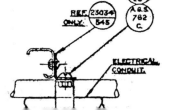

ELECTRICAL
CONDUIT.

SPITFIRE MARK I P9374

SPITFIRE MARK I
P9374

THE REMARKABLE STORY OF HOW A UNIQUE
AIRCRAFT RETURNED TO FLIGHT

ANDY SAUNDERS

GRUB STREET • LONDON

DEDICATED TO

PETER FREDERICK CAZENOVE

30 MARCH 1908 – 7 DECEMBER 1980

PILOT OF SPITFIRE P9374

Published in 2012 by
Grub Street Publishing
4 Rainham Close
London
SW11 6SS

Reprinted 2012

British Library Cataloguing in Publication Data
 Saunders, Andy.
   Spitfire Mark I P9374 : the extraordinary story of
   recovery, restoration and flight.
   1. Spitfire (Fighter plane)--History. 2. Spitfire (Fighter
   plane)--Conservation and restoration. 3. World War,
   1939-1945--Campaigns--France. 4. World War, 1939-1945--
   Aerial operations, British. 5. Great Britain. Royal Air
   Force. Squadron, No. 92.
   I. Title
   623.7'464'0288-dc22

   ISBN-13: 9781908117069

Jacket and book design by Sarah Driver

Printed and bound by MPG Ltd, Bodmin, Cornwall

Grub Street Publishing uses only FSC (Forest Stewardship Council) paper for its books

CONTENTS

	Acknowledgements	vi
	Foreword	1
	Introduction	2
1	Ghost of The Sands	4
2	Emergence	8
3	Identification	13
4	P9374 in Action	19
5	P9374's Pilots and Operational History	28
6	In Dead Men's Shoes	45
7	From Recovery to Re-Discovery	57
8	The Search	62
9	Reconstruction	76
10	The Rolls-Royce Merlin III	94
11	The De Havilland Propeller – Mission Impossible	103
12	Warpaint	113
13	Into The Air	124

Appendix

I	Spitfire I Technical Specifications	136
II	Follow-Up Test Flying	140
III	Aircraft Movement Card for P9374	143
IV	Engine History	145
V	Milestones in the History of P9374	146
VI	Paint Specifications	147
VII	All Known Recorded Flights by P9374	148
VIII	All Known Recorded Flights by Fg Off Cazenove	150
IX	All Known Victory Claims by 92 Sqn 23/24 May 1940	151
X	Two for the Future?	
	P9372	154
	P9373	158

Selected Bibliography	162
Index	164

ACKNOWLEDGEMENTS

First, I must thank Simon Marsh and all at Mark One Partners, especially Thomas Kaplan, for their enthusiastic support and encouragement. I hope it provides them with a lasting record of the story of 'their' Spitfire, and that I have managed to do justice to its extraordinary tale.

A great many friends and colleagues, as well as interested or involved parties, have provided assistance, information or photographs for this book. I am gratefully indebted to all who have gone out of their way to help. Some, I must single out for special acknowledgement.

First, my old friend Peter Arnold, the world renowned Spitfire 'guru', has been a constant and ready source of information, advice, encouragement and photographic material. His extensive knowledge of all things Spitfire related has been a real asset in this project, and he has always been willing to assist – sometimes with specific technical or historical data, but more often than not with Spitfire trivia that only someone like Peter would readily know or be able to find out. Frequently, he has fielded the most randomly off-the-wall question, and if he didn't immediately know the answer then he very soon found out. Thank you, Peter!

John Romain and all of his team at Duxford have been consistent in their help and input. Without their co-operation this book could not have been written and I must offer my sincerest thanks, and especially to Anna McDowell.

I must also single out Martin Overall and Col Pope of the Duxford team for their unstinting help and willingness to answer my many questions, as well as their generous assistance with photographs, drawings and technical data. Col Pope was especially helpful in respect of the colours and markings. Nothing was too much trouble for him and he provided me with all that I needed to know – he is surely the true 'maestro' of wartime aircraft colour schemes and their application to surviving warbirds. Thank you, Col. I hope that I have also been able to give something back to both Martin and Col, and to the P9374 project, with my own little bits of technical and historical knowledge along the way.

My friend Guy Black, who provided a management link to Mark One Partners in the re-creation of P9374, has been a source of knowledge, information and inspiration. He has also encouraged me and supported me in the production of this book. Our frequent long drives to Duxford together have often been party to long discussions about Spitfires in general, P9374 in particular and Mark I minutiae in detail. Those journeys often resulted in mutual learning and discovery of many things Spitfire. Thank you, Guy, for your valued and indeed good-humoured input.

Steve Vizard, an associate across more years than I care to think about, and his specialist Spitfire team at Airframe Assemblies on the Isle of Wight must receive special thanks from me for their input – especially Chris Michel. Thank you, Steve and Chris.

Another friend, Peter Watts of Retro Track and Air, was incredibly helpful filling in lots of detail that I needed to know about the engine and propeller. His assistance and time was always most generously and helpfully given and, as ever, he was the perfect gentleman and very patient in answering so many questions. Equally, I must thank his son Stuart for assistance with regards to the propeller construction and his help during my visit to Kemble.

Gordon Leith at the Royal Air Force Museum provided his usual prompt and friendly service and came up trumps with aircraft record cards and various photographic enquiries.

In no particular order of merit, I must also thank the following:

Winston Ramsey, Alan Hulme, Paul McMillan, Peter Cornwell, Rob Pritchard, Gerry Burke, Michael Robinson, Paul Baillie, Steve Rickards, Mark Kirby, Malcolm Pettit, Patsy Gillies, Colin Pateman, John Dibbs, Norman Franks, Chris Goss, Dennis Knight, Paul Strudwick, Edna Cazenove, Alfred Price, Tony Dyer, Sven Kindblom, Philippe Osche, Larry Hickey, David Whitworth, Bruce Rooken-Smith, Bernard Cazenove and Martin O'Brien.

All of those involved directly or indirectly in the project to restore P9374 have been tenacious and single-minded in their objective to see this aeroplane fly in its exact 1940 configuration, and precisely how it was when it was set down on Calais beach in May 1940. Without exception they are a remarkable team, and I am indebted to all of them for their help and am humbled by their truly astounding achievement. One way or another, just about all of those involved in the project – from those with the smallest role to those having a significantly major input – have all helped me with putting together this book. Thank you one and all.

With the over-arching acknowledgement above, I should have thanked everyone. However, if there is anyone I have omitted to mention individually by name, and should have done so, then please accept my sincerest apologies and gratitude for your input. There are, though, some other very important people who have been involved with this book and who I must single out.

I really want to say a special thank you to my wife Zoe who also spent a good deal of time sorting and collating facts and in putting a great deal of the detail in this book into chronological, alphabetical or numerical order for me. Your help and support was invaluable, especially over what has been a difficult period for you in your own job. I cannot thank you enough.

Last, but of course by no means least, my sincere thanks go to John Davies and his ever patient team at Grub Street, especially to Emer Hogan, Sarah Driver and Sophie Campbell. This project has sometimes moved along in fits and starts as we have tried to juggle its production schedule in concert with that of P9374. Projected completion dates and first flight dates are necessarily very moveable feasts in the world of aircraft restoration. This does not always sit very well with planned publication dates for a re-lated book and in cataloguing a book for release when the relevant dates get moved ever backwards. Consequently, and because of the various enforced delays, many have been keenly anticipating this book. Thank you for your patience. I hope it has been worth the wait!

FOREWORD

by

Sqn Ldr Geoffrey Wellum, DFC, RAF (Ret)

So much has already been said and written over the years about the Spitfire that it would at first appear to be difficult to add anything new. Some of the books, being too long and too technical, have not made for entertaining reading. Andy Saunders however has taken a different approach to the subject by focusing on the story of one particular aircraft, a Mk I Spitfire of 92 Squadron.

As a very junior newly commissioned pilot officer I was posted to the squadron towards the end of May 1940. At that time I had never seen a Spitfire, let alone flown one. Strolling down to the squadron dispersal I first cast my eyes on the squadron Spitfires and one of them must have been P9374.

To me on first acquaintance this lithe creature was a thing apart, elegant, graceful and relaxed, a thoroughbred single-seat interceptor fighter. I had never seen anything to compare with it. The day after I joined, 92 Squadron were in combat for the first time covering air fighting over the French coast leading to the evacuation of Dunkirk. By the end of the second day, in spite of many victories, five of our aircraft failed to return and one of them was P9374.

The aircraft was flown on that day by Flying Officer Peter Cazenove who crash-landed on the beach at Calais and was incarcerated as a prisoner for the remainder of the war.

The story of the salvaging from the beach at Calais where it had come down and the subsequent restoration of P9374 makes for compelling reading in this most entertaining of books, one that is a must for all devotees of the Supermarine Spitfire.

Geoffrey Wellum

INTRODUCTION

When I received information during the latter part of 1980 that a virtually complete Spitfire had been discovered on the beach near Calais I was, to put it mildly, more than a little sceptical. Having been involved in the discovery and recovery of artefacts from World War Two aeroplanes for some years, I was very well aware that relics of such aircraft were certainly to be found in abundance across the whole of Europe – and yet the possibility of a newly discovered and supposedly 'intact' Spitfire in such a public place seemed highly improbable. Indeed, the likelihood of this being anywhere near a complete Spitfire was dismissed out of hand by those in the UK who became aware of the story as no more than fanciful nonsense. However, when I got a telephone call from the manager of the Calais Hoverport I started to take a little more notice. What he told me was more than astonishing and the reports certainly now had my fullest attention! My astonishment very soon spilled over into sheer and unadulterated incredulity when he then mailed me a selection of regional Pas-de-Calais French newspaper cuttings showing recent photographs of a substantially near-complete and very early mark of Spitfire, pictured 'emerging' from a sandy beach. My informant explained that the Spitfire had appeared out of the sand not far from the hovercraft ramp and he intended to set about recovering it imminently. Since he had heard of my knowledge and involvement in such matters he wanted to know if I would be interested. He did not need to ask me twice!

Unfortunately, and by the time I was able to visit France to view the wreck, the Spitfire had been rather brutally recovered from the sands and hauled up ashore in several large sections. Sadly, it no longer looked very much like the Spitfire wreck photographed for the newspapers but on the other hand the mass of assembled wreckage was now at least wholly accessible for a close-up inspection. This would not have been the case whilst the Spitfire was buried in the sand, and thus an opportunity presented itself to search for clues as to the identity of the aeroplane and its then mysterious history. Those clues did not take long to find and unlocked a truly fascinating story.

With the identity of the Spitfire established and a name put to its former pilot there followed a degree of interest which led to the wreckage passing to the Musée de l'Air at Le Bourget, Paris, where the engine and machine guns were eventually displayed. However, the jumbled mass of material that had once formed the airframe of this Spitfire held little continued interest for the museum in terms of display material and it was not very long before the wreckage was disposed of by the museum to a private French collector and Spitfire enthusiast.

Ultimately, however, the wreckage ended up being purchased by American businessman Thomas Kaplan and returned to the UK for a full restoration to flying condition, but following the precise build specifications of the original Spitfire Mark I. To say that such a restoration was demanding, technically exacting as well as extremely costly and historically challenging would be a very considerable understatement. It is to the immense credit of all the partners involved in this ground-breaking project that the team have triumphed and seen this unique aeroplane reconstructed to her former glory and returned, at last, to the air.

Having been directly and indirectly involved in the epic of this particular Spitfire since the aeroplane's initial discovery, it was a very great honour indeed to be asked to write its history by its current owners. It is intended that this book should have the widest possible appeal across a very broad spectrum of those who are interested to learn more of this very special aeroplane, and it is a book that focusses as much on the history of the aeroplane (from 1940 to date) as on the in-depth technical or engineering details of its rebirth. I hope that I have managed to do both credit and justice to the team involved in this utterly stunning engineering project whilst, at the same time, paying tribute to that band of incredible men like the wartime pilot of this machine and others who originally flew the Spitfire in combat. In a sense, the story of this reluctant survivor epitomises the very spirit of 1940 Britain – emerging battered and scarred from battle and yet ultimately triumphant in restoration to its former glory.

Spitfire Mk I, P9374, is a truly iconic aeroplane and this is her remarkable tale.

East Sussex,
March 2012

1 GHOST OF THE SANDS

The worldwide quest for wrecks and relics of World War Two aircraft had undoubtedly gathered a great deal of momentum during the early 1970s, perhaps spurred on in part by the 1969 release of the epic film *Battle of Britain*, but also in equal measure by the enthusiasm of a new generation who had not participated in the conflict of 1939-45 and were now out-growing their Airfix kits and war comics. Suddenly, the possibility of finding and recovering tangible relics from that period opened up a whole new vista. A veritable craze for hunting aircraft wrecks sprang up around the globe, although in the UK and Europe this was mostly just a case of salvaging the often rather pathetic and battered shards from crash sites where aircraft had fallen to earth. A few were found rather more intact in coastal waters, although corrosion and the difficulty of salvage usually meant that they remained in situ. On the other hand, rather more substantial sections of aircraft could still be found on mountains and high ground in the UK – left at the places they had crashed simply because wartime salvage had proved too difficult if not impossible. Nevertheless, even these wrecks were no more than twisted hulks of partial airframes or engines and invariably in very poor condition by virtue of the nature of impact and their long exposure to the elements. Further afield, however, virtually complete wrecks were being found in jungles on Pacific islands with many such aircraft being discovered, for example, in Papua New Guinea. The enthusiasts of the UK and Europe could only look on enviously at these exciting finds on the other side of the world and content themselves with grubbing battered engines and propellers out of marsh and moorland. The likelihood of an 'intact' airframe discovery in north-west Europe seemed more than remote. All of that changed, however, during the summer of 1980.

During the early autumn of that year the author received an unexpected telephone call from a Monsieur Jean Louf who was manager of the Calais Hoverport and also a private pilot. Having heard of the author's knowledge and interest in matters appertaining to the wrecks of wartime aircraft, he wanted to know if there might be any interest in a complete Spitfire wreck which he said had been found on the beach at Calais. At first, it was difficult to comprehend the fact that what Jean Louf had described as a 'complete Spitfire' could possibly have been found there. Indeed, how could a complete Spitfire suddenly be 'found' on a French beach near Calais? Whilst this was not exactly a notable tourist region it was, nevertheless, a well visited beach close to significant centres of population. Even if it had emerged from the sea at an unusually low tide why had its presence not been known before? Surely, it would have been seen before now? At the very least someone would have been aware that it was there and word would have long ago spread in the enthusiast world. It was difficult if not impossible to comprehend that a Spitfire that apparently still very much *looked* like a Spitfire, and even with its cockpit canopy supposedly still in place, could suddenly have come to light.

Fanciful stories about such discoveries certainly surface from time to time and it seemed likely that this was *all* that had happened here; ie a *du coq á l'âne* (cock and bull) story and not a Spitfire. Having shown polite interest in Jean Louf's story, the author nevertheless agreed that he would certainly like to see the local newspaper reports about the Spitfire wreck which were to be published imminently, but fully expecting they would at best show a battered and corroded Rolls-Royce Merlin engine and perhaps a few pieces of otherwise unidentifiable airframe. Nothing could have quite prepared the author, or the historic aviation world at large, for the stunning images that ultimately appeared in the French regional newspaper *Nord Littoral* of 16 September 1980. Quite simply, and just like Jean Louf's original report, they were nigh-on unbelievable.

Under the headline: 'Out of the Sand after Forty Years' the newspaper ran the following piece of editorial, explaining that the wreck had first been sighted on the beach at Waldam, Calais, by two local beach-combers, Monsieur Barbas and Monsieur Duquenoy. Barbas said:

> "The wreckage of the fighter was about a kilometre from the Phare de Walde lighthouse just to the north of Calais. The carcass of the fighter was seen from afar with the tail, cockpit and right wing of the aeroplane clearly visible where it had emerged through the sand.
>
> "Whilst the rear tail of the plane is in pretty good shape, part of the right wing is peeling off. The left wing of the aircraft is intact and certainly still buried in the sand.
>
> "The cockpit of the aircraft was filled with sand and silt, to just below the cockpit sill. Just behind the position of the pilot's seat there is a support bracket for the radio antennae made from Bakelite and this is entirely undamaged. It is the same with the internal structures of the wings which are still present including some of the electrical circuits. The armament of the aircraft, particularly the machine gun located in the part of the wing that has emerged, has disappeared.
>
> "The engine is relatively well preserved, especially some stainless steel parts which are still intact and still protected by their original grease forty years after they went for their long swim. The engine casing itself clearly shows the name of the manufacturer – Rolls-Royce.
>
> "This indication suggests that this carcass is that of an aircraft likely to be a British Spitfire Mk IX [sic] as many of this type and mark were brought down

endu par les sables, quarante années après...

La carcasse d'un avion de chasse britannique est mise à jour à Waldam

Dans notre précédente édition, nous nous étions fait l'écho de la découverte faite sur la plage de Waldam, de la carcasse d'un avion de chasse, abattu vraisemblablement au cours de la dernière guerre.

Depuis quelques jours, en effet, de fortes marées sont enregistrées, ce qui permet à de nombreuses épaves englouties depuis plusieurs années le long de nos rivages, de revoir temporairement la lumière du jour.

Ce fut le cas la semaine dernière pour cet avion de chasse britannique, qui fut découvert par deux chasseurs, MM. Barbas et Duquenoy, deux Calaisiens demeurant respectivement rue Aristide-Briand et rue de Rabat.

Forts de leurs renseignements, nous nous sommes rendus hier matin, à la faveur de la marée basse, pour photographier la carcasse de cet avion, que le sable commence de nouveau à recouvrir.

L'APPAREIL, PROTEGE PAR LE SABLE DANS LEQUEL IL ETAIT EN FOUI EST QUASIMENT INTACT (TOUT-AU MOINS DANS SES FORMES). ON APERCOIT LE MOTEUR ET LE COCKPIT, EMERGEANT DU SABLE, L'AILE DROITE DANS LAQUELLE SE TROUVE ENCORE LES EM PLACEMENTS DE L'ARMEMENT, MAIS EGALEMENT LA QUEUE ET LA DERIVE DE L'APPAREIL (A DROITE). L'AVION A PERDU UNE DES TROIS PALES CONSTITUANT SON HELICE   Photo « Nord-Littoral »

Word of the emergence of a Spitfire from the sand reached the local newspaper *Nord Littoral*; this only served to widen public knowledge of the wreck's existence and drew yet more curious visitors to the site.

in our coastal area during the last war.

"After having been revealed to daylight thanks to the tidal action of the channel, this carcass of a Spitfire (which was last seen approximately 25 years ago) will ultimately be entirely covered by the sand once more. This is the same sand which, since 1940, has protected the Spitfire from damage by the sea."

If this editorial made for mouth-watering reading back in England, the images published with this short story certainly made for eye-popping viewing. Just as Jean Louf had described, and just as *Nord Littoral* had reported, this was indeed a pretty much intact Spitfire. Whilst only portions of airframe were protruding above the sand, those bits that could be seen were distinctively and very unmistakably Spitfire. What was perhaps more exciting, though, was that much of the airframe remained invisible and was clearly buried in the sand and was, most probably, in quite good condition.

There could be no doubting that this was certainly a significantly complete Spitfire. And it was sitting on a shoreline visible from the English coast and only a few hundred yards from the outskirts of Calais. Unfortunately, its location very close to this centre of population and its exposure in the local news media, as well as to the natural elements, were not to help its continuing intact survival over the coming weeks. The predictions of the *Nord Littoral* as to its imminent descent back into the sand may well have ended up being accurate, but it was a lamentable fact that the Spitfire would surely disappear from view somewhat substantially less intact than when mother nature had delivered it up. It was this fact, perhaps, that led Jean Louf to consider other options.

Certainly, when the Spitfire had first emerged the only visible and obvious damage already caused to the airframe had been the sawing-off of one of the propeller blades that had, at one time, protruded two or three feet above the sand. Quite possibly this had been cut off by a souvenir-hunting German soldier, although it is more likely it had been removed postwar because it presented a hazard to small marine craft and an obstruction for local fishermen. Either way, it had clearly been one blade of a three-bladed metal propeller and this certainly ruled out the suggestion by *Nord Littoral* that this was a Spitfire Mk IX. Not only were Spitfire IX propeller blades more likely to be made of wood but, more importantly, they would always have been four-bladed assemblies. This Spitfire had a *three-bladed* metal De Havilland propeller telling us that this was most likely a Mk I or a Mk V. Certainly, any number of Marks I through V had been lost in this general area – especially during the heavy fighting of 1940 and then on into the RAF Fighter Command 'Circus' offensive of 1941. For the time being, though, its precise history eluded investigators and enthusiasts.

However, at the rate the wreck was being smashed about by casual souvenir hunters there would surely be little left with which to identify its history before the Spitfire slipped back under the sands as had been predicted. Already the outboard starboard Browning .303 machine gun had been crudely hacked out of its mounting bay, resulting in the 'peeling away' of the wing described by the newspaper reports. The passage of a few more tides, and doubtless the actions of visitors standing on the Spitfire's eliptical wing-tip, eventually caused the wing to break and fail at the weakened point of the gun bay – evidence clearly visible in a series of pictures taken of the emerging wreck. Thus far, though, the plundering of the wreck had been by curious locals and casual beach walkers. In those pre-internet days word had not yet spread particularly far and wide, and it was only the author and a close circle of contacts who evidently knew about the wreck outside the immediate environs of Calais. All the same, the Prefecture of Pas de Calais as well as the local Mairie and other authorities in Calais were all singularly disinterested in the Spitfire, its protection or its longer term conservation. Whilst a rather more enlightened and protective stance might now be taken in the twenty-first century, it was different in 1980. Here was just another piece of war debris with which France was liberally endowed. It simply didn't matter that much.

Over the winter of 1980/81 the wreck showed further signs of being pillaged and was seriously deteriorating through the actions of tide and weather. The tip of the rudder had been pulled off or been washed away, ancillaries had been removed from the engine and the stainless steel 'hoops' of the sliding cockpit canopy were no longer visible and had probably been taken. Somebody had also set about the rocker box covers, smashing one of them to pieces – probably in order to remove the embossed Rolls-Royce inscription. Additionally, someone had completely removed the previously intact propeller spinner cone. Whether it was through the actions of mother nature, deliberate souvenir hunting or just wanton vandalism (or any combination of the three) the condition of this relatively intact Spitfire was degrading extremely rapidly. Jean Louf's mind was made up. There was only one course of action; recovery.

2 EMERGENCE

The recovery of any wartime aircraft wreck, and particularly a substantially complete machine like the Calais Spitfire, requires a good deal of planning and organisation in order to execute the operation efficiently and safely. Even the simplest of recoveries can end up being surprisingly complex. Logistics, legalities and a hundred-and-one other factors need to be taken into account. In fact, it is generally accepted that the recovery of a complete or near complete aircraft is often the easy part. What happens to the wreck post-recovery is invariably the challenge. Where will it be stored or exhibited? How will it be stabilised from any ongoing corrosion? How will its future conservation be addressed? Such are the questions and problems facing the RAF Museum with the potential retrieval of a near-intact Dornier 17-Z from underwater on the Goodwin Sands, even as this book is being written. Unfortunately, in the case of the Calais Spitfire, none of these factors seem to have been properly thought through. Perhaps it was just the desire to save the aircraft from further looting, or to prevent its imminent loss back to the sea, that prompted what can only be described as a hurried and rather brutal recovery that got underway on 9 January 1981.

Led by Jean Louf, a party of enthusiasts and helpers associated with the Calais Hoverport took a mechanical excavator out across the hard sand at low tide and commenced the recovery operation. The plan was simple. First, dig away some of the sand around the aircraft and remove the sand from above the port wing. Then, attach cables to the engine and propeller and simply drag the entire Spitfire out of its resting place and haul it up the beach behind the mechanical excavator. The plan was then to tow the aircraft on its belly up the concrete hovercraft ramp just a few hundred yards to the south of the crash site. Needless to say, it didn't exactly work out like that.

During its long sojurn, and when it had sunk ever deeper, the Spitfire had naturally filled completely with sand and this was now packed tightly into every internal crevice of even the parts now exposed above the sand. As a rough guide it may be considered that one cubic metre of wet sand would weigh in at between 1,922kg and 2,082kg, and even dry sand would weigh a collosal 1,602kg per cubic metre. This aeroplane, clearly, was packed with very many cubic metres of heavy sand – in the fuselage, cockpit, engine bay, wings and tail. Everywhere was stuffed tightly with the wet muddy sand and even objects and equipment within the structure such as air bottles and fuel tanks, as well as all the Spitfire's smallest fixtures and fittings had, in turn, been filled with saturated sand and water. Consequently, the all-up weight of an early mark Spitfire (around 2,400kg) was now increased by a massive factor and the strain upon an old airframe, already weakened by corrosion, was huge. Indeed, and somewhat astonishingly, each internal cubic metre of the aeroplane that was filled with sand now weighed not very far short of the all-up weight of the in-service aeroplane itself! Certainly, the strain upon a factory-fresh Spitfire hypothetically filled with *dry* sand would almost surely have been sufficient to cause significant damage and failure to its structure.

Imagine, then, the strains placed upon an already weakened airframe filled with many tonnes of wet sand. Added to that, there would have been a huge and almost incalculable element of suction from the cloying sand and mud around the Spitfire; a force that would have been unwilling easily to release its prisoner of four decades. Whilst the Spitfire *looked* and essentially *was* intact it was only thus because the wet sand inside (and thus the air-frame skinning) was being supported by the wet sand *outside*. In effect, it was only intact because it remained where it was. To remove the aeroplane and to maintain any chance that it would maintain its structural integrity and stay looking anything like a Spitfire, it was essential that the wet sand would have to be removed first. All of it.

To the author, and without even looking at any figures, it was clearly apparent that when the Spitfire first appeared on the surface, the weight problem was one that would have to be overcome if any proper salvage attempt were to be made. The difficulties had presented themselves to the author in a practical demonstration when he had only just recently helped to recover from the beach at Pevensey Bay in East Sussex just one half of one of the horizontal stabilizers (tail-planes) of a B-17 Flying Fortress that had crashed there in September 1943.

Like the Spitfire, this airframe part was also buried in the sand. Despite efforts to dig it out and move it, it still took twelve men to struggle and strain, very slowly, up the beach with the weighty object over a period of a few hours. When the compacted wet sand was finally flushed out it was found that the entire assembly could be lifted, and quite easily, by just one man and with one hand! So, with the external suction and the internal weight that had to be overcome, any prospect of recovering an intact Spitfire using the methods proposed had to be remote to say the very least. Given the problems associated with the weight of wet sand therefore, the author cautioned Jean Louf against any hasty operation to get the Spitfire off the beach. These were factors that were not taken properly into account, if at all, when the recovery did eventually begin. The consequences, unfortunately, were inevitable and catastrophic upon the continuing integrity of the airframe.

Once the strain on the cables had been taken up there was initially no movement at all from the airframe. When more power and traction were applied to the mechanical exca-vator there was, at last, some give. Slowly, something moved. But it was only the engine and propeller that had shifted forward. Unfortunately, the airframe didn't follow and with the sound of tearing metal, the crunch of rending aluminium and the noise of creaking

The recovered Rolls-Royce Merlin III engine was found to be in a surprisingly good condition; still painted black, brass data plates intact and oil in the sump.

and groaning wreckage underneath the sand, the Rolls-Royce Merlin engine tore itself free from its mounts and pulled away from the firewall as fuel, oil and hydraulic fluid spewed out of severed pipes and sheared unions cascading onto the sand.

It had been the recovery team's view that the engine formed the most substantial and robust part of the Spitfire, and to them it seemed to be the obvious (and in fact only) attachment point from which they could attempt their salvage. The possibility that the engine might shear from its mounts, or the mounts from their attachments to the airframe, had not been considered or at least it was something that had been overlooked. Now, of course, they were confronted with a still-buried but now engine-less Spitfire. The tide was coming in, and it was getting dark. It was time to retreat to the shore and consider the next move.

With publicity about the Spitfire already in full swing, and with word locally having spread very rapidly, it was inevitable that the next low tide during daylight hours would see multitudes of curious sight-seers flocking to the wreck site. Now, much more of the aircraft lay exposed whilst other previously inaccessible parts around the engine bay could finally be seen. Other loose parts lay scattered all around, dislodged as the engine was being pulled away from its mountings. Many of those parts were in very good condition, but were either subsequently pilfered or washed away and then lost by the action of the incoming tide. Clearly, the remaining wreckage could not be left where it was. Quite apart from anything else the local authorities would not wish to see the exposed and scattered metal left where it lay in situ on Waldam Plage. Consequently, Jean Louf had no option but to return next day and on the first low tide of the following morning he tried to finish off what he had started.

With nothing now substantial on which to attach any cables or hawsers, the options were

Although the previously intact Spitfire suffered badly during a rather brutal recovery, the leading edge 'D' box sections of the wings bore testimony to their surprising strength and remained intact. The red fabric patches covering the gun ports are still clearly in evidence. One of the 'Dunlop' main-wheel tyres can also be seen.

somewhat limited for the would-be Spitfire rescuers. A little bit of further digging, mostly to remove the sand and water that had flooded back into the excavations and accumulated on the overnight high tide, did very little to free up the remaining airframe in any meaningful way. If the retrieval of the engine on the previous day had been savage, the efforts to free the airframe were necessarily going to be ever more so given the methods being adopted. However, there was certainly something to be said for the robustness of the leading-edge 'D' box construction of the Spitfire's wings when wire cables were attached to the main carry-through spars and a further hauling and mauling of the airframe commenced.

Incredibly, both leading-edge sections of the wings held together and remained attached to the wing spars and internal construction in such a way that they came out as one very large lump of wreckage, and with the centre section and cockpit area being held together relatively intact. Of the trailing part of the wings aft of the D-section leading edges, however, the weight of the sand and water had quite literally pulled them apart – spilling out the guns and both main-wheel tyres as they went.

It was a similar story with the rear fuselage and tail section; the much weakened sections of skinning and flimsy portions of fin, tailplanes, rudder and elevators being dug out as just so much scrap. They no longer retained any semblance whatsoever of R J Mitchell's once sleek Supermarine creation. The last item dug from the sand was the complete tail oleo leg with its wheel and tyre. By now, only those who were intimately familiar with the Spitfire might have recognised the pile of debris as ever having been such a machine.

Once the wreckage was pulled up onto the concrete slipway of the hoverport it bore very little resemblance to the Spitfire that had first appeared in the sand. Had a more measured and considered approach been taken to its salvage, it is likely that the Spitfire could have been recovered substantially intact.

What was eventually taken up to the hoverport ramp to join the engine looked exactly what it had now become; a pile of aeronautical scrap.

In the efforts to salvage the Spitfire the aircraft had to all intents and purposes been ripped apart and destroyed as an intact although 'relic' condition aircraft. To a degree, the local fascination and perhaps that of its would-be rescuers had begun to wane once the aircraft ceased to look anything much like a Spitfire. As we shall see much later on, every cloud has a silver lining – or so the saying goes.

In the case of this Spitfire its reduction to scrap and various component parts during the rather botched 1981 recovery would ironically be the key to its longer-term survival and, ultimately, an ability to return it to its former glory and eventual flight. Had the aircraft emerged from the sand substantially intact, and if it had still retained the very appearance of a Spitfire, it is more than likely that the hulk would have been preserved as it was and thus become a significant static museum exhibit in its own right. As things would turn out, its virtual destruction upon recovery would lead to a sequence of events that would ultimately see a far more spectacular outcome for the wreck than would otherwise have been the case. Had it ended up as an inanimate museum piece then that, most likely, is where the story of this Spitfire would have concluded.

With the recovered parts now taken to the relative safety of Jean's care in the hoverport there remained, of course, the question as to what was the story behind this particular Spitfire? Who had been the pilot? What had happened to him? And how, why and when did it crash? To answer all of these questions, Jean Louf had turned in hope and expectation to the author for some specialist assistance.

IDENTIFICATION 3

Whilst it had been the author's intention to visit Calais and view the remarkable sight of the Spitfire in situ it was unfortunate that time constraints and other commitments had precluded this happening in the run-up to Christmas 1980. However, with news that recovery had actually taken place not reaching the author until *after* the event, the only purpose of a visit to Calais would be to view the recovered wreckage and hopefully identify its history. In this respect, Monsieur Louf was struggling somewhat. All he knew was that this was a Spitfire and nothing more. Of course, those who 'knew Spitfires' were already confident that the subject of interest was an early mark – its three-bladed metal propeller and a fit of four Browning .303 machine guns in each wing was evidence enough of that. So, even before anything was pulled out of the sand it was possible to determine that here

was either a Mk I or perhaps a Mk Va. At least this provided a slightly narrower time frame for the loss of this Spitfire, and it was therefore possible to say that this had most likely occurred at some time during 1940, or possibly early 1941. All the same, a great many Spitfires were lost in that period, and in that general area, and were otherwise 'unaccounted for'. As is often the way with such quests for clues it appeared, on the face of it, to be a case of a very big haystack and an extremely tiny needle.

In a telephone conversation with Jean

The recovery was masterminded by M. Jean Louf, manager of the Calais Hover-port. He is pictured here with a section of recovered control surface.

Louf on 15 January 1981 the author established, however, that the wreckage was to be moved to La Musée de l'Air at Le Bourget, Paris, on Monday 19 January and so it was a case of getting to France quickly before it was moved away from Calais. During that conversation, however, the author enquired as to any visible identification marks or clues. Were there any painted numbers to be seen? What specific mark of Merlin was the engine, and what was its number? The replies, unfortunately, were not encouraging. There were no painted numbers, although plenty of unhelpful stamped part numbers. As to the engine, this was a Merlin III and so at least this information confirmed we were dealing with a Spitfire I. That was something of a step forward, although the engine number recorded on the brass engine plate was not necessarily very helpful at this stage. The reason was simple. All RAF aircraft were allocated an aircraft movement form, or Air Ministry Form 78 to give it its proper Royal Air Force designation. These forms were specific to each aircraft, and were almost the aeronautical equivalent of a vehicle log book. On the form F.78 would be recorded key details such as the airframe number, units to which it had been allocated during its service life (and the dates thereof), any repairs and modifications to the airframe, bare details of its eventual fate and, usually, the engine number.

However, there were obvious difficulties in working backwards from just an engine number to find the aircraft serial number and thus its history. First of all, there were many hundreds of record cards for individual Spitfires lost during the period of interest and a trawl of them all at the RAF Air Historical Branch would be hugely time consuming. Secondly, there could be no guarantee that any matching number would be found because engine numbers were not always recorded on the AM Form 78 at all. More often than not at this stage they were not recorded. When they were noted, it was mostly the case that engines would have been changed whilst in-service and any replacement engine numbers were, invariably, never recorded on these forms. In effect, the engine number (if it was noted at all) was that with which the aeroplane had been delivered, and therefore unless the aeroplane had been more or less brand new at the time of its loss one would not expect to find the engine number anyway. In the absence of anything else, however, it was worth trying to find out what one could about this particular engine.

From Jean Louf we knew both numbers recorded on the engine; one of them was the Rolls-Royce number (13769) and the other its Air Ministry number (143668). It wasn't much to go on, but the records of the Rolls-Royce Heritage Trust threw up a bit of information and at least this was a start. Engine 13769 had been built at Derby on 27 October 1939, tested on 2 November 1939 and despatched to the RAF at 14 MU, Carlisle, on 6 November 1939. Unfortunately, that was about as far as the Rolls-Royce records could help and the trust could only add that it would have been fitted to either a Spitfire Mk I, Defiant Mk I, Hurricane Mk I or a Battle Mk I. We already knew that we could exclude other types and it was certainly from a Spitfire Mk I. The final detail that showed it was 'Recorded as written-off' was not particularly helpful, either. We already knew that, too! Immersion in sand and the English Channel for forty years had certainly seen to it that this was now well and truly the case. All that one could say, with any reasonable degree of certainty, was that a delivery date of 1939 on this engine perhaps pointed to a loss quite early on in 1940. There were certainly no Spitfire losses that might 'fit' during 1939, and although they were a little bit vague there was reason to suppose that the Rolls-Royce records probably indicated a loss quite early on in the life of this engine. But what else could Jean Louf tell us in advance of a visit? Was there not some other little clue that might be significant? What about the ammunition feed chutes? We knew they had been found with the guns and as they were usually marked with the aircraft number painted in red was there anything showing? There was nothing Jean could see painted, but there was a stamped number – 9578. It wasn't very much to go on, and the aircraft serial numbers of the gun chutes were always painted. Not stamped. All the same, a check was made to see if any Spitfire serial numbers could be found to match 9578 with prefix letters K, L, N, P, R and X all being checked out. Again, and not exactly unexpectedly, a blank was drawn. It was still that very big haystack. All now rested on a visit to Calais and a much more thorough inspection of the recovered remains. Perhaps that would reveal the elusive needle.

Since the wreckage was to be removed to Paris in four days, Jean Louf quickly organised tickets with Hover Lloyd for a return crossing by the author from Ramsgate to Calais on 17 January, and instructed that the tickets should be collected from a local travel agent but, most importantly and mysteriously, the recipient should report to the manager at Ramsgate upon arrival there. When the specific instructions were duly complied with it was a surprise indeed to be offered the trip over to France sitting high up in the cockpit on the jump-seat behind the pilot and co-pilot. It was an exciting start to a memorable day, and one that would ultimately prove vital in the ongoing history of this particular Spitfire. As the craft settled down onto the concrete ramp at Calais Hoverport the pile of grey corroding alloy that had once been a Spitfire could clearly be seen sitting rather forlornly and almost insignificantly on the hovercraft apron. From high up in the hovercraft cockpit it didn't look very much like it had ever been an aircraft, let alone a Spitfire! That said, the Spitfire had not after all been broken up quite as comprehensively as the author had been led to believe by Jean Louf. All the same, the author was itching to get going on a detailed inspection before the short daylight hours made the task impossible. And amongst all of this wreckage there really ought to be that vital clue. Somewhere.

The hovercraft apron held the rather more substantial 'lump' of centre section and wing leading edges with sundry other pieces piled up around and it quickly became apparent that this was a very early Mk I with its pump-up 'handraulic' undercarriage pump – a large lever located on the starboard side of the cockpit. Adjacent to this lever, and just below the longeron opposite the port-side cockpit door, there ought to have been a constructor's data plate that would have led to the aircraft serial number. Although these plates carried only the manufacturer's coded numbers these numbers can be tied, more or less, to the allocated RAF serial number. If a direct match cannot be established then it is generally possible to say within a range of one or two serial numbers what the number of

Above: The author of this book, Andy Saunders, sits in the cockpit seat of the Spitfire at Calais. His association with this aircraft would ultimately span some thirty-two years.
Below: Lying in the well of the cockpit and jammed between the rudder bars, the author found the cracked armoured glass windscreen and its frame. Retrieving the item from where it had lain for over forty years, he was gifted the item by M. Louf. It would later play a part in the future history of this particular Spitfire and its ultimate resurrection.

that aeroplane *ought* to have been. As luck would have it, only the four rivets of this plate existed – the data plate itself having either corroded away completely or having been taken long ago by an unknown souvenir hunter. More likely it was the latter, since it would have been easily accessible to visiting German troops and a tempting trophy.

Sometimes, although not always, aircraft serial numbers were pencilled if not painted onto the insides of removable access panels situated across various parts of the airframe, including engine cowlings. This was to ensure that such panels, when removed, were ultimately replaced onto the correct aircraft. Very often, the same panel or cowling would not properly fit in the corresponding location on any airframe other than the one from

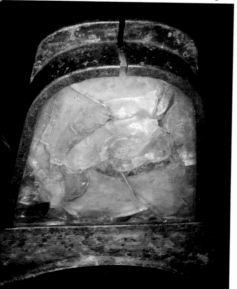

which it originated. In the case of the Calais Spitfire wreck, however, all of the internal silver paint seemed to have long ago flaked off and had been lifted away from the metal by the action of salt water, taking with it any of the vital pencil or painted numbers that may have once been present.

Thus far an exhaustive inspection of the assembled wreckage threw up no clues at all, although one or two discoveries were made that had thus far passed the salvage team by. For example, lying in the well of the cockpit and across the rudder bars and heel boards was the bullet-proof armoured windscreen, covered in silt and sand, but still held in its frame and exactly where it

Above: As the wing structure fell apart under the colossal weight of wet sand, the remaining seven Browning .303 machine guns dropped out of the wreckage. Still intact and almost pristine, this is the complete battery of four guns from the port wing. So well preserved were the guns that one was test-fired by the French army and operated perfectly after merely being cleaned and oiled!

Below: Andy Saunders interviews Jean Louf at Calais and notes down an aircraft serial number that he has just discovered faintly marked on a piece of wreckage; P9374.

had fallen when the canopy structure had finally collapsed. Fascinating though this was it helped not at all in nailing down what was thus far proving to be a very elusive identity.

Jean Louf, though, remained convinced that the engine must hold the key to the mystery and took me into the hoverport workshops to view the Rolls-Royce Merlin III and its propeller assembly. On the floor alongside was one complete battery of four .303 Browning machine guns, their ammunition boxes and the ammunition chutes. (Perhaps not surprisingly, no ammunition was on board the aircraft when the wreck was recovered and had presumably all been taken out by the Germans in 1940.)

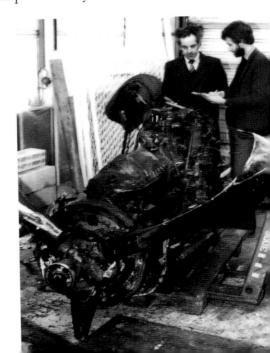

Suddenly, something was immediately apparent – the significance of which had either been overlooked or was not appreciated by Jean Louf. Very faintly, in red paint, was the inscription 'P9374' stenciled onto one of the chutes. One had to look hard, but it was very clearly there. Here at last was the aircraft serial number. It had been there, unseen, all along and would surely unlock the story behind this particular Spitfire. Incredibly, and almost before the author had finished writing the number down in his notebook,

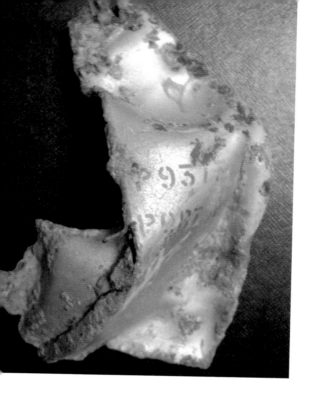

The serial number was painted in red onto one of the machine gun 'empties' chutes just like this one marked on a similar item found in the wreckage of Spitfire P9373, a 'sister' aircraft to P9374. (That Spitfire also features later in this story of P9374.)

one of the workshop engineers had taken the chute and set about it with a rotary wire brush before he could be stopped or any photographs of the vital number taken. "Voila, Monsieur!" he exclaimed, proudly holding up the gleaming gun chute which was now, unfortunately, without its all-important number! Sensing the significance of this piece to the author, the engineer had taken matters into his own hands and thoughtfully cleaned the obviously valuable item up. "Bon! Oui monsieur?" exclaimed the Frenchman. With a Gauloise cigarette dangling from his bottom lip, and a Gallic shrug, the Frenchman now seemed understandably perplexed by the apparent horror that his 'helpful' actions had caused. No matter. At least the serial number had been found and recorded, even if it no longer existed. Unfortunately, a search of the other ejector chutes revealed that all trace of the number had been obliterated by time and tide.

Overall, the saga of the recovery of the Spitfire we now know to be P9374 was described by one military history journal as 'the preservation tragedy of the decade'. Luckily, it was a tragedy that would ultimately see a complete reversal. The fortunes of Spitfire Mk I, P9374, would take a brighter turn in due course.

P9374 IN ACTION  4

Back in England it didn't take long to establish the very basic facts about the aircraft. It had been on the strength of 92 Squadron and had been lost on 24 May 1940, which proved that the initial guess that this aeroplane had been lost early on during the 1940 French campaign was correct. All that remained now was to fill in the detail of its pilot and of P9374's loss that day. To do this, a variety of sources were consulted (both published and primary source archive material) in order to try to build up a complete picture about the aircraft and the story behind its landing on Calais beach. The first port of call was the RAF Museum, Hendon, to seek out its AM Form 78. Luckily, this was very quickly traced (some having been lost or stolen since the end of the war) and even more fortunately the engine number was also recorded against P9374 and logged as 143668. This was the aeronautical equivalent of an exact DNA match, and whilst the last remaining P9374 serial number had been obliterated we at least had confirmation now through the engine number that this was indeed P9374. To an extent, Jean Louf's unshakeable belief that the engine numbers were of vital importance had been at least, in part, vindicated.

Form 78 also told us that here was a virtually new aeroplane, recorded as having flown a mere thirty-two hours and five minutes at the time of its loss. Built by Vickers Armstrong under contract number 980385/38, it had been delivered to 9 MU on 2 March 1940 before its acceptance by 92 Squadron on 6 March 1940 and having eventually been lost just over two months later on 24 May. So far the basic history of both the airframe and engine had been established, but what of the more important human story behind its pilot? Who was he, and what had happened to him on that day and since then? In order to begin unravelling this intriguing part of the story it was necessary to turn to the operations record book (ORB) for 92 Squadron.

Before examining in detail the events over Calais on the fateful day, we need to look at the background of RAF Fighter Command's operations in the airspace over this corner of France and Belgium during the spring of 1940 and the events that were about to unfold on the beaches of Dunkirk below. Operation Dynamo, the evacuation of the British Expeditionary Force and some French forces from Dunkirk, was not yet underway although, by

Above: Flying Officer Peter Frederick Cazenove, 92 Squadron, May 1940.
Below: The aluminium bucket seat from Spitfire P9374 that Fg Off Cazenove had rapidly vacated on 24 May 1940. His seat-type parachute sat in the bucket well of the seat. The structure along the seat leading edge was for the stowage of Very pistol light cartridges.

and large, the RAF's fighter and bomber squadrons that had hitherto been operating from French airfields had been, or were being, withdrawn back to the British Isles. In terms of the fighter force based in France, this had almost exclusively been made up of Hurricane squadrons.

Upon the insistence of Air Chief Marshal Hugh Dowding, the C-in-C of RAF Fighter Command, not a single squadron of Spitfires had ever been sent to French bases. Now, with air cover necessary for the evacuation and to protect other ongoing British operations on the ground in France and Belgium (operations which had now been backed right up to the Channel coast) it was, at last, possible to commit Spitfire squadrons to battle operating from airfields in the south east of England. It was during one such operation, in support of the defence of Calais and during attempts to hold back German forces from the Channel coast, that Spitfire P9374 had been lost. According to the squadron operations record book, Fg Off Cazenove force-landed at Calais and was believed to be in enemy hands (see page 54 for full extract).

Here, then, was the barest outline of the story as committed to paper in the squadron archives, although there was clearly yet far more of this tale to unravel. However, at least we now knew the name of P9374's pilot and since he had apparently seen out the duration of the conflict as a prisoner of war there surely remained a very good chance that, in 1981, he might still be living. All that remained was to attempt to track him down and, hopefully, get his own account of events that day. Maybe, and with luck, perhaps it would be possible in time to reunite him with his old Spitfire and to sit him once again in his cockpit seat. Those, at least, were the author's aspirations.

First, though, we had to find him but in 1981, which was of course a long while before the internet age, tasks like this were not as easy as they often are some thirty years or so later. With no such thing as an on-line people and address finder the hunt invariably involved a search of all UK telephone directories and although the squadron operations record book identified him as Fg Off Cazenove, rather frustratingly it did not record his initials.

However, that problem was quickly solved by reference to the June 1940 Air Force List where a Peter Frederick Cazenove was shown. It had to be the same man, reasoned the author. As a fall back, however, and in the event that no P F Cazenoves should turn up in UK directories, it had been decided to contact Messrs Cazenove, the well-known City banking institution in order to see if there was any connection. First, though, it was the telephone directory.

Although Cazenove was an uncommon name the author always operates a strict rule of thumb when searching in this way for RAF veterans; that is, look first in the *local* phone book! Time and again, with an almost monotonous regularity, those being sought ended up living right on one's doorstep, so to speak. Thus, instead of an alphabetical search starting with the phone book for Aberdeen, the local directory was consulted first and immediately a P F Cazenove was identified near Littlehampton in West Sussex. It seemed too good to be true.

After assembling all of the facts on paper, and noting down all of the questions that needed to be asked, the telephone number was duly dialled – albeit with some trepidation. Having rung for what seemed many minutes it was eventually answered by an obviously frail and elderly lady who identified herself as Edna Cazenove. Was this the right number for Peter Cazenove? Indeed it was. Unfortunately, and as Edna Cazenove explained, her husband Peter had died just a short while earlier, on 7 December 1980 – the very same day on which Jean Louf and the author were having a conversation about the mystery of the identity of the pilot and his aeroplane. It was an incredible piece of bad luck, made all the worse by Mrs Cazenove's revelation that Peter had spoken of his Spitfire not very long before he had died and posed an extraordinarily prophetic question: "I wonder if anyone will ever find my Spitfire?" he had mused. Bizarrely, and at about the same time that Peter was asking himself that very question, P9374 was just emerging from the sands and *Nord Littoral* were running their story about the discovery of the ghostly Spitfire wreck. It was certainly a great shame that Peter Cazenove never got to hear about it, or to see his Spitfire once more. Equally, it was hugely unfortunate that Peter's own first-hand account of events that day were never recorded. Almost certainly it would have been a captivating human interest story to help bring alive the tale of P9374. Instead, we have had to content ourselves with assembling the facts, so far as we can, from a multitude of other sources.

First of these sources, perhaps, ought to be the colourful account of that day's action as offered up by Larry Forrester in his book *Fly For Your Life*, the biography of R R Stanford Tuck. The next best thing to a first-hand account of that action, it is hard to read it and not ponder just how equally riveting (if not more so) would have been Peter Cazenove's own account – right up to the point that he put P9374 down on the sand and then walked into a burning Calais. Here, then, is Bob Stanford Tuck's account of 24 May 1940, as told through Larry Forrester:

"They went up and down their 'beat' twice, and then they spotted a formation of twenty Dornier 17s at about 12,000ft, far inland, heading towards the beaches

Flt Lt Robert Stanford Tuck, later to become a famous fighter ace, was one of the 92 Squadron pilots who took part in the air actions over the French coast on 24 May 1940.

to bomb the cornered troops and the evacuation fleet *[sic: the evacuation was not yet underway]*. Behind the bombers, and about 7,000ft higher – 4,000ft above the Spits – gleamed a protective arrowhead of Messerschmitt 110s. Tuck knew he had to ignore the fighters and try to break up the Dornier formation before they could start their bombing run.

" 'Buster!'

"The eight pilots pushed their throttles forward through the emergency seals to full power and the Spitfires slashed inland in a shallow, curving dive. Tuck meant to bring them round in a wide half-circle on the tails of the bomber stream, but was prepared to change direction and risk a head-on or beam attack if the escorting 110s looked like dropping down to intercept.

"A wonderful stroke of luck then relieved him of this worry. As he glanced again at the high wedge of fighters he saw a squadron of Hurricanes hurtling almost vertically down out of the remote blue, quick as minnows, and in a perfect 'bounce'. In mere seconds the Messerschmitt pack was shattered into writhing fragments, completely taken up with the business of its own survival. Now the Dorniers would have to look out for themselves.

"The bombers were flying in wide, flat 'vics' of three. As the Spits came down on their tails they were in a gentle turn to starboard, lining up to start their bombing run. Bartley's section, on the inside of the Spitfire's curving dive, got within range before Tuck's flight. As Bob recalls it, 'Tony did a rather extraordinary thing'.

" 'He went down the starboard side of the stream, shooting them up one wing, and I distinctly saw him leapfrog over one 'vic', under the next, then up over the third – and so on. He did the whole side of the formation like that, and he tumbled at least one – maybe two – as flamers at that single pass. It was just about the cheekiest bit of flying I'd ever seen. The chaps in his section tried to follow him, but they managed only one or two of the 'jumps'. Tony made every one.' "

One of those 'chaps' flying in Tony Bartley's section had been Peter Cazenove and the bombers they had intercepted were, in fact, Dornier 17-Zs of I./KG 77. Whilst the Spitfires had luckily avoided the fighter escort, the gunners on board the Dorniers were putting up a spirited defence, with Tuck describing how the gunners "blazed defiance" and "laid on a heavy crossfire". It was into this defensive crossfire that Cazenove gamely followed Tony Bartley, and although Bartley's Spitfire was hit and badly damaged it was not terminal.

Exactly what happened to P9374 we cannot be absolutely certain, although it is reasonable to assume that Cazenove's Spitfire took hits from the Dornier gunners, almost certainly disabling the engine. It only took a single round in the wrong place to cripple the fuel supply, wreck the oil pressure or to knock out the cooling system. Certainly, it would appear that nothing major from any structural point of view had affected the Spitfire and it could clearly still fly under control – albeit that Cazenove had obviously decided a return across the Channel or back home to Hornchurch was out of the question with an

Above: Another of the pilots engaged in those actions was Flt Lt Tony Bartley who was later drawn by the artist Cuthbert Orde.
Below: The enemy aircraft which 92 Squadron had been attacking were Dornier 17-Z bombers of I./KG 77 as depicted in this photograph. It is likely that return fire from one of the Dornier's 7.92mm machine guns resulted in critical damage to P9374.

engine that was either running rough or overheating. Below him lay a wide, flat and open expanse of sand with the tide far out.

Wheels up, the Spitfire skidded across the beach throwing up a great arc of sand and water as the gently wind-milling propeller kissed the surface, the blades bending under impact as the radiator and oil cooler scoops dug harshly into the wet sand. The Spitfire

finally came to a halt not very far away from the Phare de Walde light tower. Before he climbed out of the cockpit, Cazenove just had time to radio to Tony Bartley circling protectively overhead: "Tell mother I'm OK. I will be back in a few days". Turning for home, Bartley watched as his friend headed out across the beach for the distinctive grass-tufted sand dunes that mark the coast just to the north of Calais.

With a full-scale evacuation in full swing very soon after his forced-landing, and ships at that time still passing under free passage to and from French ports, Cazenove had good reason to believe that he would soon be home in order to fulfill the promise to his mother. Eventually, he made his way towards a Royal Navy destroyer – or so the story goes. Bartley takes up that story: "From his POW camp, we heard from Peter Cazenove. He tried to get on three destroyers but was turned off each one. The navy said that all the accommodation was reserved for the army, and the air force could go f*** themselves!"

From Tuck's account we also learn that Peter Cazenove had then walked up the beach towards Calais, and as he did so he

Above: Plt Off Pat Learmond. *Below:* As enemy forces eventually reached the coast, and British troops were evacuated from nearby Dunkirk, it was inevitable that P9374 would become a bit of a tourist attraction for German soldiers. Here, a party pose triumphantly with Spitfire P9374. The Phare de Walde light tower is faintly visible in the background to the left of the propeller blade.

210

Just a matter of a couple of weeks after P9374 landed on the Calais beach, the whole area was photographed by a high-flying RAF reconnaissance aircraft on 5 June 1940. This image of the Calais area takes in the beach where the Spitfire came to rest and it can actually be picked out as a dark speck (ringed) on this photograph.

chanced upon the shattered and burnt wreckage of another Spitfire. Examining the remains he found a burnt piece of parachute harness with the name 'Learmond' marked on it. He had, in fact, accidentally stumbled across the wreckage of his friend Plt Off Pat Learmond's Spitfire (P9370) shot down just the day before. Learmond, who was killed in the crash, has no known grave and is still posted as missing with his name recorded on the Runnymede Memorial to those of British and Commonwealth Air Forces who are still missing. *

Although Peter Cazenove was down and out of the fight, his war was very far from over. Meanwhile, his Spitfire became the object of curiosity for sight-seeing German soldiers as they posed for photographs with the aircraft which had already begun to settle quite markedly into the sand. Incredibly, and many years later, Spitfire specialist Peter

* NOTE: During September 1987 French enthusiasts excavated the buried wreckage of a Spitfire I close to the sea wall at Les Salines, situated between Calais and Sangatte. Amongst the wreckage had been discovered the remains of the pilot, although no formal identification of either pilot or aircraft could be made. Some researchers, however, have suggested that this may well have been the wreck of Pat Learmond and his Spitfire, P9370.

Local information pointed to the crash having taken place around 24 May 1940. If it had crashed on 24 May 1940, the only real candidate for an unaccounted for Spitfire pilot is Plt Off Richard Dennis Aubert of 74 Squadron who was shot down "south of Dunkirk" on that day .However, if the date of loss was 23 May then clearly Pat Learmond becomes a contender for the identity of this unknown pilot, and it has certainly been theorised that this is quite likely to have been the Spitfire he had been flying and his remains.

Whilst its location up against the sea wall might tend to add weight to this suggestion there are other contra-indicators to this. Given the account that Cazenove found the wreckage "as he walked up the beach" it must be recognised that the crash location at Les Salines is to the south of Calais, whereas Cazenove landed to the north. To have accessed this part of the beach then, it means that Cazenove would presumably have had to cross the harbour mouth at Calais, or else walked around the inland outskirts of the town before heading back to the shore on the south side. In some respects, this scenario seems rather unlikely, although we can reasonably conclude that Cazenove had made it into Calais port in order to attempt to secure passage on a Royal Navy vessel and he might then have walked from the port to the Les Salines location.

The only 'clue' found during the dig at the crash site was the Rolls-Royce engine plate, although its number bore no relation to that recorded for either Learmond's or Aubert's aeroplane. However ,and as explained earlier in this work, the lack of an engine number match cannot necessarily rule out either of these aircraft. The remains of this unknown RAF pilot were buried in Plot 14, Row E, Grave 22 of Terlincthun British Cemetery, Pas-de-Calais, on 28 July 1988. Just possibly, Peter Cazenove might have been able to throw some light to help solve this mystery by identifying exactly where it was he found the Spitfire wreck that he had managed to establish as Learmond's machine. Sadly, fate intervened and decreed otherwise.

Spitfire P9374 gets more visitors as it starts to settle into the sand. One German soldier looks to be intent on getting the control column top as his personal trophy. Certainly, all trace of it had long since gone by 1980.

Arnold spotted a Spitfire photograph that had been pictured with German soldiers on a sandy beach being offered for sale on an internet auction site during December 2002. Having taken part in an on-line bidding war for the photograph, Peter was fortunate enough in January 2003 to win the image of what was initially just a mystery Spitfire. The only clue, and a rather meagre one at that, was the visible top of a single aircraft identification letter which could only have been a 'J'. The possibility that this was P9374 could not be ruled out, although no other markings beyond the letter J were visible and since the individual code letter carried by P9374 was then unknown and unrecorded in the 92 Squadron archives it provided no linking evidence.

Then, however, a remarkable clue was spotted on the photograph; in the background could be seen a distinctive tower-like structure. This was the Phare de Walde light-tower, and its position relative to the Spitfire in the photograph matched *exactly* the known crash-landing location of P9374. For the first time, we were looking at a picture of P9374 *exactly* as it had been, and very soon after it had landed on the beach. It also told us that German soldiers had visited the wreck and, inevitably, had almost certainly looted a good deal of 'booty' from P9374 for personal war souvenirs.

The Spitfire was also unwittingly photographed by the RAF when a photo-reconnaissance aircraft carried out a vertical photo sweep of the entire evacuation coastline on 5 June 1940, with P9374 appearing as just a dark speck on the sand. It was an insignificant 'dot' which hid what was to be a far from insignificant story.

5 P9374'S PILOTS AND OPERATIONAL HISTORY

In her short life of thirty-two hours and five minutes flying time as an operational aircraft with 92 Squadron, Spitfire P9374 was flown by a number of squadron pilots. Clearly, she was also flown by Supermarine test pilots, both post-production and pre-delivery, and probably also for a test flight by an RAF pilot whilst waiting delivery from 9 MU, RAF Cosford, to its assigned operational squadron. Unfortunately, it has not been possible to trace any details of P9374's flights pre-delivery to 92 Squadron and only a handful of its flights post-delivery.

The following, then, is a brief summary of biographical information relating to those pilots who are known to have been directly associated with P9374, together with details of all her known and recorded flights. However, it is important to note that whilst P9374 was delivered to 92 Squadron on 6 March 1940, the operations record book for the squadron held at The National Archives, Kew, only includes entered details of the specific flights of P9374 from 9 May 1940 onwards. Generally, these flight details were logged in the Detail of Work Carried Out pages that formed part of the ORB, but in the case of 92 Squadron these pages were not routinely completed until 9 May 1940.

Thus, we have a gap in our knowledge of the individual flights of P9374 when with the squadron from 6 March to 9 May 1940. At Kew, these records are now kept on microfilm but in an effort to ensure that the original entries had not been missed during the microfilming process the author also requested and accessed the original paper document. Sadly, no trace could be found of any record for this missing period and the author is satisfied at the time of committing this book to print that they simply do not exist. That said, the log book of Bob Stanford-Tuck is still in existence and although the period when he served on 92 Squadron from 1 May 1940 was consulted it unfortunately shows no trace of this illustrious pilot ever having flown P9374.

In addition to the pilots detailed below, it must be remembered that there are certainly other pilots on 92 Squadron (in addition to test and ferry pilots) who would have flown her. It is unfortunate, however, that our list cannot be comprehensive and exhaustive.

Peter Cazenove was the principal pilot involved in our story, but it has been necessary to glean much of the information we now know from official documentation and other recorded sources, as, unfortunately, his death in 1980 left a number of gaps and unanswered questions although the author's interview with his widow, Edna, in 1981 was helpful in filling in a few of those details that otherwise would have remained unknown.

(NB: The ranks given throughout this section of the book are those relevant as at May 1940.)

Peter Cazenove was Eton educated and is seen here in a house photograph sometime during his period at the school between 1921 and 1925.

FLYING OFFICER PETER FREDERICK CAZENOVE, 73727

The Cazenove family owes its origins to the De Cazenoves who were Hugenot protestant refugees forced to flee France during the seventeenth century. Peter Cazenove was born at Broxbourne, Hertfordshire, on 30 March 1908, the only son of Percy and Beatrice Cazenove and brother to Angela and June (later Lady June Frewen, wife of Admiral Sir John Frewen). Peter's father, Percy, had served with the Buckinghamshire Light Infantry during World War One but in civilian life was a stockbroker.

Born into such a family it was perhaps inevitable that Peter would follow that career path, although this was after a public school education, first at a preparatory school in Seaford, East Sussex, and then at Eton between September 1921 and December 1925. As a junior executive in his father's company* he failed to flourish and did not enjoy the world of stock-broking and finance. Thus, it wasn't very long before he had tendered his resignation and in a complete change of career path had travelled to Argentina for his new employment with the Fray Bentos Beef Company. By 1938, though, he was back in England where he joined the Auxiliary Air Force and was commissioned as a pilot officer with 615 (County of Surrey) Squadron on 16 May of that year with the officer number 90458.

Exactly what his chosen line of work might have been upon his return to England is unclear, although his widow, Edna, stated that his pre-war employment

was as a test pilot with the De Havilland aeroplane company at Hatfield. That being the case,
one would suppose that Peter had already gained
some considerable prior flying experience, although it is not known when or where he qualified.

   As a civilian pilot he is not recorded as being
granted his RAF pilot's wings until 29 December 1938. Details of his time with 615 Squadron
are sketchy, although during that period the
squadron were based initially at Kenley and then at Old Sarum with effect from 29 August
1938. 615 Squadron were equipped first with Gloster Gauntlet aeroplanes and then with
Gloster Gadiators from 8 May 1939. Between 20 August and 3 September 1938, Peter is
shown as having attended 'general training' at RAF Thorney Island, but little else is known
of his period with 615 Squadron.

   On 24 August 1939 Peter relinquished his commission in the Auxiliary Air Force on
appointment to a commission in the Royal Air Force Volunteer Reserve, officer number
73727. However, quite apart from the momentous events that were propelling the world
ever closer to global conflict it was also a period of considerable upheaval and crisis in
Peter's personal life, too. Indeed, we find him listed in the *London Gazette* of 18 August
1939 as having been declared bankrupt with effect from 14 August. Interestingly, his occupation at this time is shown as 'stock broker's clerk' with addresses at Lowndes Square,
Knightsbridge, and Sloane Avenue, Chelsea. There is no indication of any occupation
with De Havillands and it would therefore seem likely that he had returned to his father's
business – albeit with apparently disastrous consequences.

   It is possible, although by no means certain, that his bankrupt status had required him
to resign his commission with the AAF and the corresponding timings of his relinquishment of commission and declaration of bankruptcy might seem to point to this. Indeed,
the RAF King's Regulations of the period states: "An officer who has not been guilty of
misconduct may at any time be called upon to retire, or resign his commission, should the
circumstances in the opinion of the Air Council, require it." Quite possibly the 'circumstances' of bankruptcy might well have required this course of action, although that does
not exactly explain the apparently immediate granting to him of a commission within the
RAFVR, unless of course the volunteer reserve were rather more tolerant of such things
or inclined to turn a blind eye!

   Either way, it is certainly the case that attitudes towards bankruptcy were rather less
benevolent then than perhaps they might be in today's enlightened world. Of course, it is
also possible that the relinquishment of commission brought about by bankruptcy might
well have already been in train, because the proceedings were opened in June 1939 and,
as a result, this might have been unstoppable. Additionally, and with the situation between

Britain and Germany deteriorating significantly in late August 1939 to the point that war was clearly inevitable, maybe the view of the Air Ministry towards such things was rather more sanguine – hence, perhaps, his re-engagement upon commission on 24 August of that year, and just a little over a week before the declaration of war.

As if the ordeal of bankruptcy were not enough, Peter was injured playing rugby almost on the outbreak of war (he had played in the First XV at Eton) when he fractured his elbow and shoulder blade. As a consequence of this he spent September and October on sick leave, before joining his first posting as a flying instructor at No 1 Elementary Flying Training School. Interestingly, No 1 EFTS was based at Hatfield and was previously designated No 1 Elementary and Reserve Flying Training School which had been operated by the De Havilland Aeroplane Company, Stag Lane, Hatfield. It is entirely possible, therefore, that this is the De Havilland connection to which Edna Cazenove had alluded. Again, these are all details that could so usefully have been clarified by Peter had it not been for his death in 1980. However, it is recorded that on 16 November 1939 Peter was promoted to flying officer whilst at No 1 EFTS. Having miserably sat out the beginning of the war, first of all injured and then as an instructor, he must have been delighted to

Before being let loose on a Spitfire, Cazenove was checked out on a Miles Master dual-control trainer like this one. Although this is not the Master attached to 92 Squadron, it is likely that the unidentified pilots in this photograph are preparing for the same sort of flight.

receive a posting to a front-line fighter unit, 92 Squadron, on 18 March 1940. Not only that, but it was newly equipped with Spitfires.

Peter almost certainly spent the initial few days finding his feet on his first operational squadron, and no doubt familiarising himself with the Spitfire, reading the pilots notes, looking over the aeroplanes, talking to other squadron pilots, getting the 'feel' of the cockpit, where all the instruments and controls were situated and understanding the start-up procedure. It was not until 23 March that his first flight is recorded thus in the operations record book:

Squadron Leader Roger Bushell, the enigmatic CO of 92 Squadron during May 1940. (He also features later on in this account of Spitfire P9374.)

"Low cloud at first. The Commanding Officer arrived at Croydon in a Spitfire and gave dual in the Master (N7534; author) to Fg Off Cazenove. The Commanding Officer then took the Master to Gatwick. Fg Off Cazenove went solo on Spitfires."

The CO, Sqn Ldr Roger Bushell, had no doubt wanted to check out Peter's competence and confidence as a pilot in the Miles Master before sending him off solo on a Spitfire. Peter would have been given some local flying experience and been shown the flying field from the air, no doubt allowing him to visualise (and practice) his eventual take-off, approach and landing in the Spitfire. As with any pilot, it must have been a memorable day for him on 23 March 1940 when he first took a Spitfire up – notwithstanding the matter-of-fact one line dismissal of the event in the squadron record book. It would, of course, be rather nice to say that his first Spitfire flight was in P9374 although, unfortunately, the records do not exist to be able to confirm which particular Spitfire he might have first got his hands on. However, and since that aircraft had been delivered on 6 March the possibility that it was in fact P9374 cannot be ruled out. As we shall see, we know that Peter very regularly (although not exclusively) flew P9374 during his short time as an operational pilot with 92 Squadron.

The first recorded *detail* we have of Peter flying a Spitfire is on 9 May when he is shown as flying N3194 on a flight from Northolt to Croydon between 14.45 and 15.40 hours, and the first recorded flight Peter made in P9374 was apparently on 11 May when he flew a sector reconnaissance flight of fifty minutes, taking off at 10.15hrs. (*A table of all recorded flights made by Peter Cazenove in P9374 and other Spitfires is shown in Appendix VIII along with a table of all recorded flights by Spitfire P9374 in Appendix VII*)

As we have already seen, P9374 was most likely hit in the engine or coolant system just before Peter Cazenove put the aircraft down on the beach at Calais on 24 May, although it is probably fair to say that this episode was very much just the start of his adventures. After being turned away from possible evacuation by ship, he made his way into a burning and shattered Calais where he is said to have taken part in the desperate rearguard defence

of the town after having joined up with the Queen Victoria Rifles. Sheltering against bombardment in a ruined cellar, family legend has it that he was quaffing a glass of champagne when a bullet shattered the glass as he raised it to his lips. Moments later he heard those infamous words: "Hande hoch! For you the war is over!"

Initially, Peter was marched with other prisoners to the football stadium at Desvres which the Germans were using as a temporary POW holding point before transferring the captives onwards to established camps.

This is the only known photograph of Spitfire P9374 in 92 Squadron service. The squadron codes (GR) and the individual code letter J, are visible although we know that the letters GR were later removed. (See Chapter 12)

At first, and as with all prisoners of war, he was simply listed as 'missing' and although it is not known when he was confirmed as prisoner of war through the Red Cross he is shown as missing in a published list dated 13 June, albeit that the squadron operations record book for 24 May 1940 stated "…it is supposed that he is now in enemy hands". In his postwar MI9 de-brief, Cazenove reveals that he attempted to escape no less than three times during the forced-march from Calais to Desvres, but was re-captured each time. One of his companions on that march was Flt Lt W P F Treacy, a Spitfire pilot with 74 Squadron who had also been shot down and captured on 24 May. Luckier than Cazenove, Treacy eventually made a clean break and achieved a 'home run', managing to reach England in the following year.

From Desvres, Cazenove went first to Trier, then through various holding and processing camps before ending up in Oflag IIA at Prenzlau. From here, he was transferred to Stalag Luft I at Barth on 20 August 1940 where he was promoted, *in absentia*, to Flt Lt on 5 January 1941. He stayed at Barth until 1943, and in that year was transferred to Stalag Luft III at Sagan in Poland where he met up with his old CO, Roger Bushell, who was also famously incarcerated there.

Here, at Stalag Luft III, preparations were already under way for the construction of tunnels 'Tom', 'Dick' and 'Harry', to be utilised in what later became known as the Great Escape. It is understood that Cazenove was involved in the preparation of forged paperwork, although the bulk of his large frame (he measured well over 6ft and had very much a rugby player's build) had prevented him from working in the confined space of the tunnels themselves. Forged papers had already been something of his forte, having put his apparent skills in this area to good use in Stalag Luft I where he managed to walk out of the main gate on forged documents – albeit that he was re-captured shortly afterwards. According to Edna Cazenove, Peter had also been denied a place in the tunnel break-out itself because of his large stature.

Given the fact that fifty of the re-captured escapees were executed by the Germans it was perhaps fortunate that Peter didn't make the break given the selection criteria evidently applied by their captors in later deciding who should be shot and who should be spared. Indeed, one of those shot was Cazenove's commanding officer on 92 Squadron, Roger Bushell. Known as 'Big X' in Stalag Luft III, it is also highly likely that the unfor-

tunate Bushell had one time piloted P9374. The murder of Roger Bushell had a profound effect upon Peter Cazenove who had served under 'Big X' operationally, and had also endured the privations of captivity and the extreme mental and physical stress of preparation for the mass break-out with the highly respected and often larger-than-life RAF officer.

Little else of any great substance emerges from Peter Cazenove's MI9 de-brief, although he states his pro-

During his postwar period in Kenya, Peter Cazenove was an enthusiastic Polo player with the Sergoit Club. He is seen here (right) in his Polo rig circa 1954.

fession to be 'Agriculture' and also notes that he studied for his national diploma in agriculture whilst a prisoner. He also comments that he had applied to be allowed to volunteer for farm work at the same time, but somewhat tetchily reported that he had been refused permission. Given his record of prior escape attempts it was perhaps not a surprising outcome to his request. Interestingly, under the space for entering 'Squadron' he has written 92 but then crossed this through and put 615 instead, whilst under the heading 'Command' he has written AAF. The reasons for this are unclear.

Although his formal date of release from the RAF is uncertain, Peter married Jean Martineau at St Saviour's Church, Upper Chelsea, on 8 November 1945 and the couple then moved to Nairobi, Kenya, to work for Messrs Dalgettys. Here, he also ran a small farm where he enjoyed the relative high-life of a white settler farmer and indulged his passion for playing polo with the Sergoit Polo Club. Unfortunately, the white African farmer's idyll was rudely disrupted when Peter became caught up in the horrors of the Mau Mau uprising in that country, along with most of the other white settlers there.

Above: Peter Cazenove's medal awards included (left to right) the 1939-1945 Star, Air Crew Europe Star, War Medal and Air Efficiency Award.
Below: The edge engraving of Cazenove's name on the Air Efficiency Award.

Ultimately, his marriage to Jean did not survive and the couple divorced, with Peter returning to England and Jean re-marrying to become Jean Aschan. Peter then married Edna Hollis in 1956, and again moved back to Africa (this time to Ghana) as an agricultural representative. Eventually going back to England the couple embarked upon various property business enterprises and although Peter yearned to be a farmer once more, ill-health brought about by a lifetime of exceptionally heavy smoking finally forced his retirement to Angmering-on-Sea near Littlehampton in West Sussex. It was here that he died from emphysema on Sunday 7 December 1980.

FLYING OFFICER CHARLES PATRICK GREEN, 90134

Charles Patrick Green (left), was born at Pietermaritzburg, Natal, South Africa, on 30 March 1914. After graduating at Cambridge (where he was also elected a Fellow of The Royal Geographic Society) at the end of 1935, Charles decided to spend a year travelling in the United States of America. After seeing the film China Clipper with his friend Billy Fiske the pair decided to take up flying. (Fiske would later become the first American citizen to die in World War Two, whilst serving with an RAF fighter squadron during 1940.) They lost no time in their venture, and the very next morning had commenced flying lessons in a Fleet bi-plane trainer.

'Paddy' (as he was later known in the service) joined the Royal Auxiliary Air Force in 1936 and in the same year won a bronze medal as part of the Winter Olympics four-man bobsleigh team but at St Moritz in 1937 he broke a leg skiing in the President's Cup. That same year he was posted to 601 (County of London) Squadron on 28 February, and on the same day he made his first flight with Flt Lt Roger Bushell in a Hawker Hart (K2970). He later qualified as a day pilot in the Hawker Hart and as a night pilot in the Hawker Demon and was awarded his flying badge with effect from 19 September 1937. On detachment to 79 Squadron he flew the Gloster Gauntlet II biplane fighter and when 601 Squadron re-equipped with Gauntlets in November 1938 he again flew this type. In March 1939 he trained on a twin-engined Airspeed Oxford in preparation for conversion to Bristol Blenheim fighters with which 601 were being equipped. He first went solo on a Blenheim on 19 March 1939, flying L6617. In April his flight commander Bushell, allotted him Blenheim L6618, coded UF-M, to share with another pilot, Guy Branch, and at the end of the month 'Paddy' had 300 hours of flying to his credit.

With war imminent the squadron was mobilised on 23 August 1939, although by 20 October 1939 Charles Green had been posted to 92 Squadron at Tangmere to command A Flight, and where his old friend from 601 Squadron, now Sqn Ldr Roger Bushell, had also just been posted as commanding officer. At this stage, 92 Squadron were also equipped with Blenheims, although they were about to transition to Spitfires.

He continued flying Blenheims until February 1940, by which time he had 450 flying hours and then went on to Harvards (and Miles Masters for air experience) preparatory for transition to the new fighter. He finally received ground instruction on a Spitfire which was jacked up in a hangar. He flew his first solo on a Spitfire I (P9368) on 8 March 1940 and on 19 May 1940 he is recorded as having flown P9374 from Northolt to Hendon in a twenty-minute flight.

Green continued to fly Spitfires with 92 Squadron until 23 May when the squadron moved to Hornchurch, and on that same day he flew an offensive patrol to Calais/Dunkirk/Boulogne in N3167, GR-G, on what was to be the first engagement with the enemy for 92 Squadron. Although Green claimed a 'probable' and a 'damaged' (both Messerschmitt 109s), the squadron's Plt Off 'Pat' Learmond was lost. At 1.45 the squadron took off on another patrol over the same area, with Green again flying N3167.

He damaged a Messerschmitt 110 on this operational patrol, but Sgt Paul Klipsch was shot down and killed and Sqn Ldr Roger Bushell and Flying Officer John Gillies were both shot down and taken prisoner of war. Paddy Green was hit in this engagement and was badly wounded by a bullet which lodged in his thigh. He flew back across the Channel in a daze, faint from loss of blood with his thumb stuck in the wound and pressing on the severed arteries, and with the oxygen turned full up to stop him from fainting. Green managed to fly back to the nearest airfield at RAF Hawkinge, where the medical officer saved his leg but forecast a long spell in hospital.

The medical officer was right, for there ensued long periods in hospitals at Shorncliffe, Dutton Homestall, and Roehampton, which was followed by a period of sick leave. It was 10 October, 1940, before Paddy flew to Biggin Hill to command 421 Flight at Gravesend. This flight was formed out of 66 Squadron for the specific task of carrying out patrols to intercept any aircraft coming in over the coast. These operations were given the name Jim Crows.

On 11 October, Paddy flew a Hurricane, and the next day was again wounded while piloting a Spitfire (one of the first Mark IIs in 11 Group). He fought to escape from the cockpit all the way down from 18,000 feet to 1,000 when he eventually made it out of the sorely stricken aeroplane. His wounds were not so severe this time so he was flying again by 1 November 1940. On the 15th the flight moved to Hawkinge from West Malling where it had been since 30 October.

On 25 November 1940, Paddy Green destroyed a twin-engined Dornier 17 and *The Times* newspaper reported: "A Spitfire pilot, flying at 5,000 feet first saw the raider 7,000 feet above him when 15 miles off Dover. He climbed to attack and fired two short bursts from below and astern at 350 yards range. Breaking away he climbed again and dived on the enemy. Pieces fell from the Dornier, and the RAF pilot saw the enemy machine losing height." The kill was confirmed by the Royal Navy as having fallen into the sea off Cap Gris Nez.

On 5 December 1940, Paddy claimed another probable, this time a Messerschmitt 109, and on 27 December 1940, he shared a Dornier 215 as destroyed with another pilot.

On 11 January 1941, 421 Flight became 91 Squadron with Paddy Green now as a squadron leader and CO. He was awarded the DFC on 3 April 1941 but in June was posted to HQ Fighter Command as a staff officer. A strange claim to fame is that in October 1941 he flew a captured Heinkel 111 bomber for the film First of the Few. On 13

October Paddy was attached to 54 OTU at Church Fenton before being posted to 600 Squadron at Predannack on 14 November 1941 to command A Flight flying Beaufighters.

On 2 June he was posted as a wing commander to command 125 Squadron at Fairwood Common. By September he had notched up 1,000 flying hours but on Christmas Eve 1942 he flew from Hendon to Maison Blanche in Algiers to command 600 Squadron with effect from 10 December 1942. Green's score remained at two destroyed, two probables and two damaged until 5 May when he damaged a Junkers 88.

On 9 July 1943 he flew Lt Gen (later Field Marshal) Montgomery and General 'Boy' Browning to inspect the 1st Airborne Division. With the invasion of Sicily starting the next day he was again in action to some purpose, destroying a Ju 88 and a He 111 on 12 July. The Squadron score for that night was six and Paddy's tally by the end of the month was nine destroyed, two probables and four damaged. He was awarded the DSO on 29 July.

On 11 August he claimed another Ju 88, and on 6 and 7 August he had flown with Air Marshal Sir Arthur 'Mary' Coningham as a passenger. On 9 September he claimed a probable He 111, blaming his shooting for not having made it a confirmed. On 25 January 1944, he shot down a Ju 88 and on 6 March was posted to command 1 MORU (Mobile Operations Room Unit) with the rank of group captain.

On 8 July 1944, Paddy received the American Distinguished Flying Cross and he also later received the Russian decoration, the Order of Patriotic War 1st Class. On 5 November 1944, he was posted to command 232 Wing with his last logged service flight being made on 12 August 1946, his last two years of service having been spent with the Central Fighter Establishment. He returned to South Africa at the close of his air force career, having flown over 1,800 hours on fifty different types of aircraft during ten years service. He had eleven enemy aircraft confirmed as destroyed, three probables and four damaged.

Charles Patrick Green died in May 1999 and was arguably the most illustrious of all the pilots who had flown P9374.

SGT STANLEY MICHAEL BARRACLOUGH, 66487

Stanley Michael Barraclough (known as Michael) was born on 2 April 1917 in Livingstone, Northern Rhodesia, although at a young age he returned to England to live with his maternal grandparents in Shoreham, West Sussex, before moving to Chichester. Here, Michael (left) attended primary and senior schools and excelled at team sports although during the late 1920s came news that Michael's father, a colonial civil servant, had vanished in the bush. No trace of him was ever found.

Michael, having decided upon a career in the Royal Navy, joined the training ship TS *Mercury* but in 1931 he watched the Schneider Trophy seaplane races and his mind was set; a career in aviation was, after all, the path for him. In 1932 he became an RAF apprentice metal rigger (service number

566190) at Halton and upon completion of training was posted variously to Cranwell and then Iraq before applying for flying training as a pilot. His application was duly approved and he began his training in early 1939 before finally passing out as a pilot and being posted to 92 Squadron on 3 March 1940, just at the end of the squadron being equipped with Blenheims. On this type he completed sixteen hours and forty minutes flying time before converting to Spitfires. He is known to have flown P9374 on 15 May and 19 May 1940, each occasion being a night flying test.

Having served with 92 Squadron through the testing and trying period of May 1940, with the catastrophic actions over the French beaches and English Channel, he served with the squadron during the Battle of Britain but was then posted to 7 Operational Training Unit, Hawarden, as an instructor before a further posting to the Central Flying School, Netheravon, and thence to 15 Flying Training School. He was commissioned as a pilot officer on 28 June 1941 although remained in an instructor role and also became an airborne interception radar instructor on Mosquito aircraft. He was subsequently promoted to flying officer (28 June 1942) and then to flight lieutenant (28 June 1943).

Postwar, he was given the task of re-forming 504 Squadron at Syerston, Notts, in May 1946 and was promoted to squadron leader on 1 August 1947. Various staff officer posts ensued and he also graduated at the RAF Staff College, although in 1952 he resumed flying duties as deputy chief instructor with 231 Operational Conversion Unit flying Canberra bombers and based at RAF Bassingbourn. He retired from the RAF on 28 May 1958 after twenty-five years of service, and then went to work in exports and administration before moving to East Anglia. He died on 25 April 2006.

PILOT OFFICER JOHN SAMUEL BRYSON, 41823

John Bryson (left) was born on 18 November 1912 in Westmount, Quebec, Canada and after attending the local primary school he went to Westmount High School in 1926 where he was a hard working student and where he also excelled at maths. As a resident of Quebec it was natural that he should also speak French and it was a language in which he was fluent. By 1929 he had been enrolled at St Albans private boarding school in Brockville, Ontario, and by the age of fifteen he had become a strapping lad of six feet two inches with a powerful physique. On 25 June 1935 he enrolled with the Royal Canadian Military Police (the 'Mounties') and served in a number of outlying and remote posts as well as Montreal.

On 3 October 1938 he bought himself out of the RCMP and travelled to England to gain a short service commission in the RAF where he learnt to fly at 13 Flying Training School, Drem, firstly on Tiger Moths before gaining twin-engine experience on Avro Ansons and Airspeed Oxfords. This twin-engine experience resulted in a posting to a Blenheim squadron and he found himself joining 92 Squadron at RAF Tangmere on 21 October 1939. His physical bulk quickly earned him the nickname of 'Butch' and it is said that Bryson and Cazenove, both of them very big men, found the Spitfire cockpit extremely restrictive once the squadron had converted to that type – especially with the cockpit canopy closed. Even with the rudder pedals wound fully forward both pilots found leg room an issue.

Bryson is known to have flown P9374 on 10 May 1940 (practice attacks), 14 May (sector recce) and again on 18 May for another sector recce. Once 92 Squadron came back into the fray after resting at Duxford, Bryson was again back in action. Over Dunkirk on 2 June 1940 he damaged a Heinkel 111 and on 24 July he shared in the destruction of a Junkers 88 which crashed on Martinhoe Common, near Lynton.

John 'Butch' Bryson was shot down and killed in Spitfire X4037 near RAF North Weald on 24 September 1940. He was still serving with 92 Squadron at the time of his death and he lies buried in St Andrew's Churchyard, North Weald Basset, Essex. He was 27 years old.

SGT PETER RAOUL EYLES, 56889

Peter Eyles (left) was born in 1916 at Llandaff, Cardiff, Wales although by 1924 the family had moved to Falmouth and Peter was educated at Truro School. On leaving he joined the RAF as an apprentice instrument maker in the Electrical & Wireless School on 10 January 1932 and was first posted to RAF Henlow and then Habbaniya, Iraq, on 8 January 1935. Recommended for pilot training, he was eventually posted to 11 Flying Training School, RAF Shawbury, on 15 April 1939 before going to 92 Squadron on 23 October 1939. Having been a keen sailor and swimmer he earned himself the nickname 'Sailor'. Whilst serving with 92 Squadron he is known to have flown P9374 for a night flying test on 17 May 1940.

Although Eyles did not take part in the battles over the French beaches on 23 and 24 May 1940, he flew his first operational flight (escorting Blenheims to Calais) on 25 May in Spitfire P9434. As it happened, the bombers were pounding German forces attacking the port as Peter Cazenove sheltered in a cellar there!

On 2 June 1940 he shared in the destruction of a Heinkel 111 over Dunkirk with 'Butch' Bryson and then claimed a Heinkel 111 destroyed on 11 September 1940. Unfortunately, he was shot down into the sea off Dungeness on 20 September in Spitfire N3248, an aeroplane he had regularly flown since May 1940. No trace of him was ever found and he is remembered on Panel 14 of the Runnymede Memorial.

PILOT OFFICER CECIL HENRY SAUNDERS, 42893

Cecil Saunders (left) was born on 7 July 1911 in Forest Hill, London, and attended Whitgift Middle School, Croydon, between 1924 and 1930 where he excelled academically and at sports. It was at Whitgift that a master gave him the nickname 'Sammy', a name that would stick and remain with him throughout his RAF career.

He joined the RAF on a short service commission on 24 August 1939 and was sent to a civil flying training school at Derby before going to 14 Flying Training School at RAF Kinloss. On completion of his flying training he was posted to 92 Squadron on 20 April 1940. He is known to have flown P9374 on 16 May for circuits and landings, and again on 19 May when he ferried the Spitfire the short hop from Hendon to Northolt after it had gone 'unserviceable' for a projected escort flight to Paris. Like Cazenove and Bryson, Saunders was another exceptionally tall pilot who struggled within the confines of a Spitfire with his six foot three and a half inch frame.

On 4 July 'Sammy' Saunders shared in the destruction of a Heinkel 111 but was hit and shot down in a combat over Rye, East Sussex, on 9 September 1940. This was the same combat in which Plt Off Watling of 92 Squadron was shot down in Spitfire P9372 and Saunders crashed not far from Watling's Spitfire, coming to rest in a crash-landing at Midley and being admitted to a RAMC hospital at nearby Brookland, Kent, with shrapnel wounds in his leg.

Returning to operations in October 1940, he shot down a Messerschmitt 109 on the 29th of that month and claimed a Junkers 87 on 1 November and damaged a Messerschmitt 110. In that action his Spitfire (X4555) was also damaged and he crash-landed three miles east of RAF Eastchurch, unhurt. On 1 December he claimed another Messerschmitt 109 and on 5 February 1941 he shared in the destruction of another Junkers 87 near RAF Manston. This was his last action with 92 Squadron.

In May 1941 he was posted as a flight commander to 74 Squadron at RAF Gravesend and on 27 June claimed a probable Messerschmitt 109. The squadron left for the Middle East in April 1942 but Saunders then joined 145 Squadron in the Western Desert in late July 1942. On 3 August he claimed a further probable Messerschmitt 109 and on 11 September he destroyed an Italian Mc 202, before damaging another Messerschmitt 109 on 22 October and claiming another as destroyed on 25 October.

Awarded the DFC on 4 December 1942, he was then posted to 71 Operational Training Unit at Abu Sueir, Egypt, but by July 1943 was involved in the invasion of Sicily where he acted as fighter controller on board a fighter direction ship. He then took command of 154 Squadron at Lentini East in August before the squadron moved to Italy and then in February 1944 to Corsica in order to cover the American invasion of southern France. He continued to command the squadron until October 1944 before going back to 145 Squadron as commanding officer in July 1945. He retired from the RAF with the rank of wing commander on 5 May 1958 and died of a pulmonary embolism on 1 September 1992.

SGT RONALD HENRY FOKES, 88439

Ronald Fokes (left) was born in 1913 at Rawmarsh, Yorkshire, although his family later moved to Surrey where he entered Hampton Grammar School in 1925. Above average as a scholar, a keen horseman and an enthusiastic shot, Ronald joined the RAF Volunteer Reserve on 10 April 1937. During the course of his RAFVR service, Ronald (or 'Ronnie' as he was now universally known in the RAF) trained with 5 Elementary & Reserve Flying Training School, Hanworth, although he was later given a posting to 151 Squadron (Hurricanes) on 6 March 1939 and then to 87 Squadron (another Hurricane squadron) on 24 March before returning to 10 E & RFTS, Yatesbury. Called up for full-time service on 1 September 1939, Fokes went first to 1 Initial Training Wing, Bexhill-on-Sea on 14 December and then to 92 Squadron on 15 January 1940. He is known to have flown P9374 on 9 May 1940 for fifty-five minutes.

Fokes was another pilot of 92 Squadron who was involved in the destruction of a Heinkel 111 over Dunkirk on 2 June 1940, also claiming two more as probable. On 4 July he shared in the destruction of yet another and also shared in the destruction of a Dornier 17 on 10 September then damaging another Dornier on 15 September. His score of claims during the Battle of Britain continued to rise: Junkers 88 (probable) 24 September, Messerschmitt 109 (probable) 30 September, a Heinkel 111 and two Messerschmitt 109s (confirmed) 15 October, Messerschmitt 109 (confirmed) 26 October, a Junkers 88 (shared) 9 November, Messerschmitt 109 (destroyed) 15 November and a Messerschmitt 109 (probable) on 17 November. By any standards it was an impressive score.

Ronnie Fokes continued to increase his score of victory tallies, and on 15 November 1940 he was awarded the DFM and then promoted to pilot officer on 29 November. He maintained his run of victory claims before being 'rested' from operations and a posting to 53 OTU, Heston, on 1 May 1941 as an instructor and then a transfer to 61 OTU, still at Heston. In November 1941 he was posted to 154 Squadron as a flight commander until March

1942 when he went to 56 Squadron, newly equipped with the Hawker Typhoon at RAF Snailwell. In August of that year, having gained considerable experience on the type, he was seconded to Gloster Aircraft Ltd as a Typhoon test pilot before going back to operations with another Typhoon unit, 193 Squadron, in March 1943. In July of that year, he was promoted to squadron leader and command of 257 Squadron (another Typhoon squadron).

On 12 June 1944, six days after D-Day, 257 Squadron were attacking targets around Caen when his Typhoon was hit by flak and he was forced to bale out. He landed, badly wounded, behind enemy lines and was taken to a German field hospital at Chateau-de-St Loup before he succumbed to his grievous wounds on 16 June 1944 having flown operationally in the Battle of France, Battle of Britain and Normandy. He lies buried in the Commonwealth War Graves Commission plot at Banneville-la-Campagne, Normandy, France.

PILOT OFFICER DESMOND GORDON WILLIAMS, 41890

Born on 12 July 1920 at the family home of Marcle Court Farm, Little Marcle, Herefordshire, Desmond (left) grew up in Jersey when the family moved there in 1930. After an education at Victoria College, Jersey, the academic and sporting 'Bill' Williams joined the RAF in January 1939 on a short service commission. His initial flying training was undertaken at the civilian flying school at Hanworth, before he was posted to 11 Flying Training School at RAF Shawbury. Here, Williams trained on Airspeed Oxford aircraft although he was involved in a night-flying accident on 10 July 1939 when his Oxford crashed on landing. He was shaken but otherwise unhurt. On 21 October 1939 he was posted to the (then) Blenheim-equipped 92 Squadron, his twin-engine experience on Oxfords suiting him to operational flying on the Blenheim.

On 23 May 1940, one day before the loss of the aircraft, Williams flew P9374 on the Calais/Boulogne/Dunkirk patrol and 'blooded' the aircraft in combat (See Chapter Six). On that occasion he flew P9374, operationally, for one hour and forty-five minutes.

Williams flew with 92 Squadron during the Battle of Britain, and achieved a final total victory score of five enemy aircraft destroyed, one shared destroyed, one unconfirmed destroyed, one shared unconfirmed destroyed, two probably destroyed and six damaged. To this list, however, may be added one further damaged. This was a Dornier 17-Z that he and Fg Off J F Drummond (also 92 Squadron) had attacked over Brighton on 10 October. Subsequently, however, Williams (in Spitfire X4038) collided with the aircraft flown by Drummond (Spitfire R6616) and both of the pilots were killed. Williams crashed to his death at Fallowfield Crescent, Hove, whilst Drummond baled out with bullet wounds to an arm and a leg. Unfortunately, he was too low for his parachute to deploy and he fell to his death as his Spitfire crashed at Jubilee Field, Portslade. Subsequent research has identified that the Dornier 17 they had been attacking crash-landed in France with a dead air gunner on board.

Desmond Williams' family had fled their home in Jersey just prior to the German occupation of the Channel Islands and eventually set up a new home in Salisbury, Wiltshire, and it was to here that his body was taken for burial in London Road Cemetery, Salisbury.

Of those we know to have flown P9374, it will be noted that most achieved some ultimate success as fighter pilots. Unfortunately, Peter Cazenove's career as a fighter pilot was cut short in untimely fashion. Had it not been for the fact that he was forced down on 24 May 1940 and taken prisoner of war there is no reason to doubt that he would not have served with distinction as a fighter pilot during the Battle of Britain, and possibly beyond.

NOTE: There may well have been other flights by pilots on P9374 which have gone unrecorded.

These, then, are the men who flew P9374. Despite the fact that this book is essentially the story of that aeroplane, it is surely the men who flew it and who are no longer with us who should be remembered each time 'their' Spitfire is seen in the air. They could have no finer tribute or memorial to their achievements as fighter pilots in Britain's darkest hour than the re-creation of Spitfire P9374 and its return to flight.

IN DEAD MEN'S SHOES

6

Already we have seen some intimation in the preceding chapters that things did not go at all well for 92 Squadron during its baptism of fire on 23 and 24 May 1940. In fact, they went very badly. And yet, for all that, the squadron operations record book for 23 May 1940, 92 Squadron's first day of action, described the events as 'a glorious day'. The next day, 24 May, and with things not a whole lot better for the squadron, the record book is significantly less fulsome in its enthusiastic reaction to what occurred although it casually passes off its casualties that day (including Fg Off Cazenove and P9374) with an almost dismissive 'we only had two losses'. It was almost as if that was an acceptable attrition rate, but given all that the squadron had endured the day before it was the straw that broke the camel's back. On 25 May the squadron had a new CO and was out of the fray to re-form, re-equip and receive new pilots. The 'real' war for 92 Squadron had also been a real shock. Let us first examine events immediately leading up to 23 May, and then the day itself.

In the run-up to 92 Squadron's involvement there had, of course, been an actual shoot-ing war going on for other squadrons of RAF Fighter Command. Already, across France and the low countries, Hurricane squadrons had been heavily engaged with the enemy since the launch of the German 'Blitzkreig' on 10 May. Now, with the allied army's retreat to the Channel coast in full-spate, the fighting had come much nearer to and within range of home-based RAF fighter squadrons. Up to now, no Spitfire squadrons had been com-mitted to the campaign in France and all Spitfires had been carefully husbanded by Dowd-ing, the C-in-C, for home defence. Now, in late May, the evacuation from Dunkirk was set in motion and those home-based Spitfire squadrons were committed to the fight.

Unlike many of their contemporaries in French-based Hurricane squadrons, however, the Spitfire boys were greenhorns when it came to battle. They had doubtless tried to listen and learn from what had been going on in France, but they had no real concept of the brutal reality of fighting the Luftwaffe. For the Hurricane boys it had been a tough hands-on experience and they had lived rough and fought hard. Back in England, the somewhat cosseted Spitfire pilots had been insulated from such an experience in their comfortable quarters from where they had enjoyed evenings off to go up to London with their fast cars and to party with their glamorous girlfriends and carefree chums. The daytime flying routine had not changed much from how it had been in peacetime, but all of that was to change very quickly – and with a vengeance.

92 Squadron Spitfire refuelling during the winter of 1939/40 or early spring of 1940.

Whilst the squadron was nominally on more of a war footing from 31 March 1940, when it had just about fully received its complement of Spitfires, life continued much as before with a routine of battle-climb practices, practice interceptions, night flying and formation flying from the squadron's Croydon base. Meanwhile, however, the squadron was now divided into five sections; one section at readiness, three available and one released in preparedness for action. So this routine continued up until 10 May, when the sudden change in the situation on the ground in France altered everything, with the 92 Squadron operations record book noting: "Owing to the progress of the war in France today everyone was recalled from leave and all leave has been stopped."

For the next few days, however, it was very much 'business as usual', and although the war was moving ever closer the nearest that 92 Squadron had thus far got to it was when sections of three Spitfires at a time flew escort missions for a Flamingo aircraft to and from France on 15, 16, 19, 21 and 22 May. At least two of these operations though (on 16 and 22 May) had been to escort Prime Minister Winston Churchill to Paris to meet with Reynaud, Daladier and Gamelin and other French leaders. Here, at Quai d'Orsay, on 16 May Churchill learned that the situation was "incomparably worse than we had imagined". It was information that would catapult 92 Squadron ever sooner into the fray, with the shocking news imparted to Churchill that the Germans were expected in Paris in a few days at the most, and although neither Fg Off Peter Cazenove nor Spitfire P9374 were actively involved in those escorts there is good reason to suppose that P9374 was scheduled to participate in the escort flight that took place on 19 May 1940.

In addition, this planned escort flight for P9374 provides us with a probable clue as to why this aircraft, when it landed on Calais beach on 24 May, carried only its individual identity letter, J. At this time, the squadron identification letters were GR, and thus one would have expected that P9374 would have worn the full set of letters; GR-J. (*NB: a number of published sources quote the squadron codes for 92 Squadron at this time to have been*

Pilot Officer Allan Wright climbs out of his 92 Squadron Spitfire, GR-S, N3250, after combat on 2 June 1940. Ground crew are already inspecting bullet holes in the Spitfire's tail. Wright was also involved in the air actions over the French coast on 23 and 24 May 1940.

QJ although there is no doubting that at this time they were GR, the change to QJ coming some little while later.) That P9374 did not wear the full set of codes is almost certainly explained by the fact that those Spitfires involved in escort flights to Paris had their squadron identifying letters obliterated for security reasons.

What we do know is that P9374 went from Northolt (where the squadron was now based) to Hendon from where the escort to France would be flown. P9374 was flown to Hendon at 07.05hrs by Flt Lt P C Green, in company with Plt Off A C Bartley (P9373) and Sgt R Fokes (P9367). However, one can only conclude that P9374 was 'snagged' due to some technical hitch and at 08.10hrs Plt Off Saunders ferried N3193 over to Hendon as a replacement, although the problems with P9374 cannot have been too serious as Saunders then ferried the Spitfire back to Northolt at 08.40hrs. It could have been something as simple as an inoperable radio or a jammed cockpit canopy cover that had resulted in P9374 being pulled from the flight line.

Already, though, P9374 had been readied for the escort operation and in preparation her GR codes had been painted out; a seemingly insignificant event but one that provides us with one important clue showing why P9374 was devoid of her squadron markings when she later arrived at Calais. Meanwhile, Flt Lt Green took N3193 on the escort flight during the morning of 19 May, returning later that afternoon. P9374's experience of France was yet to come.

This historic back-story against which the to-ings and fro-ings of 92 Squadron, and indeed of P9374, was being played out would ultimately become a very significant part of the story of the aircraft itself and of its eventually hapless pilot, Fg Off Cazenove, just a few days later. This was the end-game in France, with capitulation and withdrawal from Dunkirk just around the corner and Winston Churchill organising and marshalling his ground forces in their now desperate plight. For the next few days, however, life continued much as it had for the pilots of 92 Squadron; operational practice flights and an otherwise general routine. New pilots, though, continued to be posted in, with Plt Off T S 'Wimpey' Wade and Plt Off G H A Wellum arriving on 21 May, all but on the very eve of battle. Both pilots would later make a mark for themselves on the squadron, but would not take part in what were to be impending actions over the French channel coast.

It was probably with a mixture of envy and relief, then, that the two new boys watched the entire squadron as it departed Northolt at dawn on 23 May to fly down to its forward operating base at Hornchurch, just that little bit closer to the action now concentrating on the ground inland from Calais and sweeping around towards Dunkirk and Gravelines. Ordered off at 10.45hrs *(NB: according to the squadron operations record book which then confusingly times the ensuing combat being at about 08.30)*, the twelve 92 Squadron Spitfires roared purposefully out towards their ordered patrol line of the Calais, Boulogne and Dunkirk coastline. Not with them, however, on that first combat operation were either Fg Off Cazenove or Spitfire P9374. The rest of the squadron, though, were headed for a truly bloody initiation although the operations record book saw it in a rather more positive light:

> "At about 08.30 hours they ran into six Messerschmitts and a dog fight ensued. The result was a great victory for 92 Squadron and all six German machines, Me 109s, were brought down with only one loss to us. It is with the greatest regret that we lost Plt Off P A G Learmond in this fight. He was seen to come down in flames over Dunkirk *[sic]*."

In fact, from a postwar analysis of Luftwaffe losses it would appear that 92 Squadron had engaged with the Messerschmitt 109s of 1./JG27 and had most likely shot down Ofw W Ahrens (POW) and Fw A Potzsch (POW), with Ahrens possibly falling to the guns of Flt Lt R R S Tuck and Potzsch to Plt Off Bryson. However, there appears to be no trace of other losses that might match the other four claimed Messerschmitts, leading to the conclusion that 92 Squadron had probably over claimed. Despite the loss of Learmond, it had perhaps been a *reasonably* successful encounter in terms of loss:claim ratios, particularly in view of the fact that this was 92 Squadron's first contact with the Luftwaffe. Notwithstanding that the squadron *thought* they had accounted for six Messerschmitts, but had probably only got two, it could have been a far worse blooding.

As the day went on, however, things would get more than a little worse for the squadron. However, it is important now to put into context the events that were unfolding on the ground, since these were shortly to become inextricably linked with the fortunes and misfortunes of Fg Off Cazenove and P9374.

When Churchill had gone to his 92 Squadron-escorted meeting in Paris on 16 May it was clear that the position on the ground was already terminal for the allies, but by 20 May the German advance across France had curled westwards out towards Amiens and along the Somme towards the sea and thence, beyond Abbeville, up the coast northwards past Etaples, Boulogne and onwards in the direction of Calais and Dunkirk. With the bulk of the British army falling back towards Dunkirk, and with plans in place for the evacuation, it became apparent to Churchill that Calais must be held at all costs. Later, he would set out his thinking:

> "Some days earlier *[than 23/24 May:author]* I had placed the conduct of the defence of the Channel ports directly under the Chief of the Imperial General Staff, with whom I was in constant touch. I now resolved that Calais should be fought to the death, and no evacuation by sea could be allowed to the garrison, which consisted of one battalion of the Rifle Brigade, one of the 60th Rifles, the Queen Victoria Rifles, the 229th Anti-tank Battery RA and a battalion of the Royal Tank Regiment with twenty-one light and twenty-seven cruiser tanks, and an equal number of Frenchmen. It was painful thus to sacrifice these splendid, trained troops, of which we had so few, for the doubtful advantage of gaining two perhaps three days, and the unknown uses that could be made of these days."

This, then, was Churcill's later summary of those desperate days during the defence of Calais, but in typical Churchill-esque style he sent the following signal to Brigadier Nicholson, who was commanding the defence of Calais, at around 2pm on 25 May: "The eyes of the Empire are upon the defence of Calais, and His Majesty's Government are confident that you and your gallant regiment will perform an exploit worthy of the British name."

In simple terms, it was an 'England expects…' moment at the time of a desperate do-or-die rearguard defence. So serious was it that even outdated and obsolescent Hawker Hector biplane bombers, fit only for the scrap heap or for pilot training in some quiet backwater, were thrown into a dive-bombing attack against the Calais assailants. As it happened, some of 92 Squadron's pilots had trained on Hector aircraft, and they had been obsolete even then. Ultimately, of course, Calais would fall – but for the moment the Spitfires of 92 Squadron patrolled and did battle in the skies above. Down on the ground the beleaguered defenders were, before very long, to be joined by one of 92 Squadron's pilots. For now, however, we return to the events of 23 May.

Having got back at around lunchtime, the excited chatter in the Hornchurch messes must have been about the action of that morning, albeit somewhat tempered by the loss of Pat Learmond. Although only officially 'missing', the other squadron pilots who saw him go down were doubtless certain of his fate. Squadron Leader Bushell must have surely de-briefed his men, but before long was probably briefing them for another squadron sortie with take-off time at 17.20 hours. Filling the slot left vacant by Learmond, Fg Off Gillies took his place. It was very much a case of dead men's shoes.

Again, it was back to the same patrol line although the pilots peering down from their cockpits at the smudges of smoke and random explosions on the ground could have had little grasp of how desperate things really were. Calais, on Churchill's orders, was 'fighting to the death', and a desperate British and French army were, effectively, in full retreat up to the channel coast across much of northern France. In the air, too, the Germans were more than notable by their presence and it wasn't very long before 92 Squadron were once more engaging the enemy. Again, the squadron operations record book told how it was:

> "In the afternoon the squadron went out again on patrol and this time encountered at least forty Messerschmitts flying in close formation. The result of this fight was that another seventeen German machines (Me 110s) were brought down and 92 Squadron lost Sqn Ldr R J Bushell – the Commanding Officer – Fg Off J Gillies and Sgt P Klipsch*. Flt Lt C P Green was wounded in the leg and is now in hospital at Shorncliffe. The remainder of the squadron returned to Hornchurch badly 'shot-up' with seven Spitfires unserviceable."

Sgt Klipsch (left) had been shot down and killed in that action, crashing to earth at Wierre Effroy in the Pas-de-Calais, with Fg Off Gillies being shot down in the same area and taken prisoner of war. The commanding officer, Sqn Ldr Bushell, crash-landed his damaged Spitfire in open countryside to the east of Boulogne. Later, from a POW camp, Bushell sent his report of that action:

> "I was shot down by Messerschmitt 110s, but managed to get two of them first. As soon as the battle started about four or five of the Messerschmitts fell on me and, oh, boy, did I start dodging! My first I got with a full deflection shot underneath. He went down in a long glide with his port engine pouring smoke. I went into a spin as two others were firing at me from my aft quarter. I only did one turn of the spin and pulled out left and up. I then saw a Messerschmitt below me and trying to fire up at me, so I went head on at him, and he came head on at me. We were both firing, and everything was red flashes. I know I killed the pilot, because suddenly he pulled right up at me and missed me by inches. I went over the top of him and as I turned I saw him rear right up in a stall and go down with his engine smoking. I hadn't got long to watch, but he was out of control and half on his back. My engine was badly shot-up and

The 92 Squadron CO, Squadron Leader Roger Bushell, was shot down in Spitfire N3194, GR-Z, just to the east of Boulogne during the squadron's first day of action. Bushell set fire to his Spitfire (seen here) and was subsequently taken POW, only to be murdered by the Gestapo on 29 March 1944 after being re-captured following his lead role in the Great Escape.

caught fire. My machine was pouring glycol. I don't quite know what happened, but I turned things off and was out of control for a while but got straight at about 5,000 feet.

"I shut everything off and the fire went out and I glided down. There was a lot of glycol, and I could not see anything much, but I turned the petrol on again and tried to make Saint-Inglevert aerodrome. The engine ran for a little while and then everything seized and a lot of smoke and fumes came into the cockpit. I reckoned it was time to bale out, and opened and undid everything. However, there appeared to be no fire, so I decided to land, undercarriage up. This I did successfully and only took a knock on the nose. When I hit, the old girl burst into flames, and, as you can imagine, I moved pretty quickly.

"I landed just to the east of Boulogne and, of course, imagined I had come down in friendly territory. The machine was blazing, but I had a look at it and could see some pretty hefty holes.

"After I landed I sat by my machine and when a motor-bike came down the

* NOTE: Sgt Paul Klipsch was lost in Spitfire P9373. The wreckage of this aircraft has also been acquired by Mark One Partners, the owners of P9374. Further details relevant to P9373 may be found in Appendix X on page 158)
Sqn Ldr Bushell was flying Spitfire N3194, GR-Z, and was probably shot down by Fw Langenburg of 5./ZG 76. Bushell had the misfortune to land in enemy-held territory, just as the Germans were sweeping northwards up the coast in their push towards Calais.

road I thought it was French. It wasn't, and there was nothing to be done about it. Thereafter, I had a long journey here *(to a POW camp; author)*. This is an air force camp where we are treated very well indeed."

During that same action, however, P9374 was finally 'blooded' at the hands of its pilot, Plt Off Desmond Williams, who was in the thick of the fighting. He gave the detail of that action in his combat report:

"Green Section, of which I was number two, was lower than the rest of the squadron. The enemy formation split and I saw three Me 110s below the main formation. On the first enemy aircraft I delivered a beam attack, changing to a quarter attack. After a five second burst I saw bits fly off the wings and one of the engines gave off clouds of white smoke. I then saw a second Me 110 in front of me, broke off from the first, and delivered a dead astern attack on the second enemy aircraft, firing three bursts of three seconds each at 200 yards range. I saw bits flying off the wings of the enemy aircraft, and a little smoke issuing from the engines. I then saw a third Me 110 spiralling towards the ground. I got on its tail and he started a very steep climb. I followed it, still on its tail, firing short bursts of two to three seconds until my ammunition was expended. I saw bits flying off both wings and noticed that the starboard engine was dead. I followed the aircraft to the ground and saw it crash."

The morning's tally claimed by 92 Squadron had been over-optimistic, but the afternoon's 'bag' was even more so. The Messerschmitt 110s encountered by the squadron were from ZG76, and although three of that unit's aircraft were reported damaged and their crews wounded there were no actual losses, with all three aircraft noted as damaged but re-pairable. *(Note: It is therefore impossible to reconcile Plt Off Dudley Williams' definite claim that he saw an Me 110 hit the ground and crash given the known Luftwaffe losses in action that day.)* Thus, the postwar analysis of the actual Luftwaffe losses in the action where 92 Squadron claimed an astonishing seventeen Me 110s destroyed cannot in any way substantiate that claim.

Yes, the squadron had certainly engaged and attacked the Messerschmitts, with at least some damage reported as being inflicted by Pilot Officers Bartley, Williams and Holland. But 92 Squadron had certainly not knocked down almost a dozen and a half of the enemy as they had supposed. Moreover, another three pilots were now lost, including the CO along with Pat Learmond's 'replacement', and the majority of the returning Spitfires were so shot about that they were now unserviceable. It had, in reality, surely been a rout and not a victory of any sorts, although it did not stop the somewhat gung-ho author of the operations record book going on to note:

NOTE: That "glorious day", however, had also seen the involvement of another Spitfire later linked to the story of P9374; namely, P9372. The wreckage of P9372 has been acquired by Spitfire owner and restorer Peter Monk for the basis of a rebuild to flying condition. The full story of P9372 is covered in Appendix X on page 154.

"It has been a glorious day for the squadron, with twenty-three German machines brought down, but the loss of the Commanding Officer and the three others has been a very severe blow to all, and to the squadron which was created and trained last October by our late Squadron Leader."

In two days of fighting, the squadron had effectively become non-operational although it was yet another full twenty-four hours before the unit was declared so and noted to be 'resting'. By some miracle, the fitters, riggers and armourers had turned around and patched up sufficient Spitfires overnight on 23/24 May to achieve eight aeroplanes ready and on the flight line by dawn. Of course, there was also an ever-dwindling band of available squadron pilots. Only the day before, Gillies had stepped into Learmond's shoes and he, too, had failed to come home. All the while, Fg Off Peter Cazenove had watched impatiently from the side lines as his pals flew off and into battle but it must have now been with a mixture of trepidation and excitement that he found himself on the roster to make what would be his first operational flight.

At 08.05 hours on the morning of 24 May 1940, Cazenove took off from Hornchurch in P9374 and headed out with the seven other Spitfires on what was already a familiar patrol line for the other squadron pilots; Calais, Boulogne and Dunkirk. Dead men's shoes had yesterday been a curse for Gillies, and now

A Spitfire I is re-armed after returning from an air battle.

Cazenove was effectively taking the place of that unlucky pilot. This thought must surely have been at the back of Cazenove's mind as the smoking ruins along the French coast hove into view. By now, the squadron diarist recording events in the operations record book was becoming adept at presenting 92 Squadron's war in a very matter-of-fact tone, rather as though he were reporting nothing more significant than a dull and routine day. It all seems quite unimportant, and with a description of the day's uninteresting weather taking up almost as much space as the description of Cazenove's loss! Here is the relevant extract:

> "Another patrol was carried out at 08.30 hours from Hornchurch and another seven Messerschmitts were attacked and brought down. This time, we had only two losses, indeed one was only a slight casualty. Flight Lieutenant Tuck* was slightly wounded in the leg, but is able to continue the good work. Flying Officer Cazenove force-landed on the land [sic: possibly a typographical error for 'sand'] and it is supposed that he is now in enemy hands. Showers of rain during the day."

New boy Geoffrey Wellum, not yet having flown a Spitfire and desperately frustrated at being left behind as his squadron went off to war, wrote of his thoughts after the remnants of 92 Squadron came straggling back home from those first encounters with the enemy over the French coast:

> "Never shall I forget the sight of those pilots as they came into the mess that

* NOTE: Although he confuses dates and timings, and mistakenly dates the episode when he is wounded as being on 24 May in Larry Forrester's *Fly For Your Life*, the account of Tuck being wounded is again worth quoting:

> "Almost out of ammunition and very low on fuel, Tuck turned for home, crossing the Channel at about 500 feet. A throbbing in his right leg reminded him he'd been hit. He found his thigh was sticky with blood and did his best to staunch it with a handkerchief. He didn't feel giddy or sick, so he told himself it couldn't be serious."

After he had landed at Hornchurch, the account of his wound continues:

> "Tuck's leg was very stiff now, and when he got out of the cockpit he couldn't quite straighten it. He noticed a small tear near his right trouser pocket, felt inside and fished out from his loose change a pocked and buckled penny: this coin had stopped one bullet, but another must have lodged in the back of the thigh."

Ultimately, it would turn out that the object embedded in the back of his right thigh was not a bullet but a small duralumin nut from the rudder pedal. It had been smashed off by the passing bullet that had hit the penny, and then embedded into his thigh through the force of impact. Famously, the Medical Officer remarked to Tuck that had the wound been just a few inches higher then he would have had to transfer him into the Women's Auxiliary Air Force! Tuck kept the penny for many years as a lucky charm and he was photographed with it for a 1960s Sunday newspaper supplement, although the whereabouts of this famous coin is now unknown.

first evening at Hornchurch after two days of hard fighting. The CO, my big blonde flight commander Paddy Green and Pat Learmond were not among them. Pat Learmond was seen to go down in flames and Paddy badly wounded. Nothing has been heard of the others as yet. It gave me my first intimation of what war is all about. These pilots were no longer young men with little care in the world, they were older mature men. On that day alone, most of them had been over Dunkirk three times and it was a day of the very fiercest fighting. No quarter asked and none given. They now know fear. They know what it takes to conquer fear because they have done it, not once but three times. Having been through the mill as these chaps have been, and having conquered fear, that most common of human emotions, can anybody ever be quite the same again?"

For Peter Cazenove, though, there had been just the one fear-filled operational flight. Just as soon as it had begun, so Cazenove's career as a fighter pilot was over.

By now, and from previous chapters, we know about Peter Cazenove's landing on the beach, but in this chapter the author has presented a background picture to that single event. Had Cazenove landed further northwards up the coast beyond Gravelines, and towards Dunkirk, then the outcome for him, if not for P9374, might well have been different; in all probability he might well have eventually made it aboard one of the evacuation ships. That said, a good number of downed RAF pilots were turned off the rescue vessels on the grounds that all available spaces were reserved for soldiers and there is an account, as we have seen, that suggests Cazenove had already been turned off Royal Navy ships that might have returned him to England, although the evacuation was not yet in operation.

To an extent, those RAF men who were ultimately not allowed to board vessels were probably victims of the widespread perception that the RAF had 'stayed at home' during the evacuation period, and were also suffering the general hostility towards the air force that was being directed by army personnel. However, the perception of the army was very far from the reality of the matter – and the experiences of 92 Squadron over France, and of many other fighter squadrons, bears adequate testimony to that. For Peter Cazenove, though, there was to be no evacuation from Dunkirk or from anywhere else. When he walked into Calais he unwittingly wandered into the last-ditch defence of that town which Churchill had exhorted should be fought 'to the death', although by 25 May Anthony Eden had told the War Cabinet that he was moving the British Expeditionary Force to the Channel coast for evacuation.

During the afternoon of 26 May, Calais finally fell to overwhelming German might and the town was surrendered at around 16.00 hours. That same day, the battered remnants of 92 Squadron had retired north to lick their wounds and to re-group and re-equip under their new commanding officer, Sqn Ldr Sanders. At the very moment Calais was falling to the Germans and Peter Cazenove was being taken into what would be a long captivity, the once boisterous but now shaken survivors of 92 Squadron gathered in the Duxford mess, silently over cups of tea, in order to be introduced to their new CO. The squadron operations record book reported thus: "As the squadron is now resting at Duxford there is nothing to report. The weather continues fair with occasional heavy showers."

Later, the Air Officer Commanding 11 Group, RAF Fighter Command, sent the following signal to 92 Squadron:

> "Air Officer Commanding sends congratulations to No.92 Squadron in their magnificent fighting and success in the first day of war operations and sincerely hopes that the Squadron Commander and the other two missing pilots will turn up later, as many others have in the past fortnight."

They were to be forlorn hopes.

Meanwhile, on that same day, Winston Churchill had asked the Chiefs of Staff to consider whether Britain could fight on alone against Germany and Italy, and whether they could 'hold out reasonable hopes of preventing serious invasion'. May 1940 had seen the worst ever military defeats for the allies and the desperate state that 92 Squadron had found itself in was merely a reflection of the parlous state of the nation as a whole and its fighting readiness. As P9374 sank beneath the sand and waves, Britain also came perilously close to going under.

FROM RECOVERY TO RE-DISCOVERY

7

As we have seen earlier, the initial interest and excitement over the discovery of P9374 in 1981 had quite rapidly waned in the historic aviation world of the period, especially when the aircraft had ceased to resemble a fighting Spitfire after rescue from the sand.

At first, the wreckage – including engines and guns – was transferred to the Musée de l'Air at Le Bourget where it was exhibited for a while but then also spent another period in outside storage. With little or no conservation or preservation attention applied to the wreckage, it began to degrade further. Indeed, exposure for decades to salt water had already taken its toll due to a lack of any inhibition of the set-in corrosion and this lack of care and attention, as well as outside storage for some of the material, was obviously having a detrimental effect. Ultimately, and perhaps fortunately, the wreckage was disposed of by the museum to a French enthusiast, the late Jean Frelaut.

A Spitfire enthusiast, he was working on the restoration of Seafire L.F.IIIc, PP972, and items from amongst the wreck of P9374 were sacrificed from that aircraft to enhance the build of PP972. At this stage there was probably little appreciation of the fact that what remained of P9374 might yet form the basis of any viable Spitfire re-build project.

The recovered wings of Spitfire P9374, abandoned in an outside storage area at Le Bourget during 1981.

Above: The Rolls-Royce Merlin III engine from Fg Off Cazenove's Spitfire also suffered a similar although temporary fate during its time at Le Bourget.

That it was recognised as having such potential during the early part of the twenty-first century can be attributed to a renewed desire to see more Spitfires return to the air.

With the passage of time, it is simply the case that the on-going value of such 'starter-kits' is of necessity decreasing in quality as the years progress. The world has now been stripped and scoured of all such available airframes from museums, gate-guard duties, scrapyards, jungles or kibbutz playgrounds! Today, those attempting to embark upon an airworthy Spitfire project need to set their sights much lower in terms of the start point – as witness the cases of P9372 and P9373 elsewhere in this book. In fact, during the late 1980s or even 1990s it is unlikely that the sad remains of P9374 would have ever been considered as a candidate for reconstruction. For that reason, Jean Frelaut naturally saw

Below: Jean Frelaut's Seafire restoration project (PP972) in France became the recipient of 'donor' parts from Spitfire P9374 after Frelaut acquired the remaining wreckage from the Musée de l'Air sometime prior to September 1981.

Peter Monk's Spitfire IX, TA805 (known as 'The Kent Spitfire'), was the recipient of the tail-wheel strut from P9374 during its re-build. It is still flying today with the same tail strut.

the butchered P9374 merely as a spares source for his Seafire project and never as having any potential as a 'restorable' aircraft in its own right.

When Frelaut had stripped and taken what he needed, the remains were in an ever-more sorry state and their value, probably, considered by many as nothing more than a collection of historic curios. Even in that category, it seemed likely that the long-term survival of the remnants of P9374 (arguably historic relics of the Battle of France in their own right) was not exactly safely assured. Thankfully, those in the world of Spitfire preservation tend to keep an eagle eye on any developments worldwide. Thus it was that when American entrepreneur Thomas Kaplan expressed an interest in owning a unique and early mark of Spitfire there was one such in 2000 that presented a possibility. Of course, and as we have seen, there had been degradation of the extant remains from the moment in 1980 when the Spitfire had been recovered right up until it left the estate of the late Jean Frelaut. Apart from what Frelaut had used, other parts had been sold on or exchanged with other owners and restorers in the intervening years. A case in point being the tail-wheel strut assembly which has since been acquired by Spitfire owner and restorer Peter Monk and is now flying on his Spitfire IX, TA805.

Ultimately, the remaining parts were acquired by Simon Marsh and Thomas Kaplan

CERTIFICATE NUMBER G-MKIA/R3

1 Nationality and Registration Marks	2 Constructor and Constructor's Designation of Aircraft	3 Aircraft Serial Number
G-MKIA	VICKERS SUPERMARINE LTD SPITFIRE 1A	6S-30565

4&5 Name and Address of Registered Owner or Charterer

MARK ONE PARTNERS LLC
C/O BUTTERFIELD BANK (CAYMAN) LTD
68 FORT STREET, PO BOX 705
GRAND CAYMAN
KY1-1107
CAYMAN ISLANDS

6 It is hereby certified that the above described aircraft has been duly entered on the United Kingdom Register in accordance with the Convention on International Civil Aviation dated 7 December 1944, and with the Air Navigation Order 2009.

Sue Wood
For the Civil Aviation Authority
Aircraft Registration
CAA House
45-59 Kingsway
London WC2B 6TE
Tel 020 7453 6666
Fax 020 7453 9670
E-Mail aircraft.reg@caa.co.uk

DATE OF ISSUE 13 AUGUST 2008 10:46 UTC+1

NOTES (a) The person in whose name an aircraft is registered may or may not be its legal owner. Prospective purchasers are warned, therefore, that this Certificate of Registration is not proof of legal ownership.
(b) No entries or endorsements may be made to this Certificate except by the Civil Aviation Authority

SEE FURTHER NOTES OVERLEAF

The registration document for Spitfire P9374 (see page 157 for explanation of the 6S–30565 serial number.)

who had them removed to the facility of Airframe Assemblies at Sandown Airport on the Isle of Wight on 14 October 2000. Here, under the watchful eye of Steve Vizard and his team, the aircraft remains could be surveyed and assessed. Whilst it was confirmed that significant parts that had been extant in 1980 were sadly no longer present, it was adjudged that what remained was more than sufficient material on which to base an ambitious build project. Certainly, there was a great deal more present with the wreckage of P9374 than had been the case with a number of other recently commenced projects. Not only that, but there was a clearly evidenced provenance trail that established more than robustly the identity of these remains as P9374. With a decision taken ultimately to proceed with getting P9374 back into the air, she was registered to Simon Marsh under the UK civil aircraft registration scheme on 16 November 2000 as G-MKIA – a highly appropriate registration in view of the circumstances!

After acquisition by Marsh and Kaplan, the parts of this historic Spitfire remained on the Isle of Wight until 19 March 2002 when they were transferred to storage with Craig Charleston's aircraft restoration workshops in Essex. Not unusually with projects of this type and nature, nothing evident and tangible occurred to take forward the reconstruction process for another few years although work was going on behind the scenes. Indeed, quite apart from any initial survey and feasibility study a great deal of research, study and investigation was required in order to plan the projected reconstruction and to work out how such an ambitious venture should be organised and project-managed. Whilst progress was slow, the aircraft was re-registered to the same owners (by now trading as Spitfire Partners LLC) on 3 March 2005 and still as G-MKIA.

With the project gathering some momentum, and with Thomas Kaplan now single minded in his intent to get P9374 flying, the remnants were again moved from storage; this time from Charleston Aviation Services to Historic Flying Ltd at Duxford where they arrived during August 2007. By now, Historic Flying had been contracted to build the wings of P9374 and Airframe Assemblies, to build the fuselage. The Rolls-Royce Merlin III engine, meanwhile, was placed in the skilled and capable hands of Retro Track & Air at Dursley in Somerset, a company with extensive Rolls-Royce Merlin engine experience and currently the company contracted to the Ministry of Defence to look after all of the engines of the Battle of Britain Memorial Flight fleet. The propeller is another story – but that was also placed with Retro Track & Air as a specialised project which is dealt with in Chapter Eleven.

On 8 July 2008 the completed fuselage of P9374 was delivered to Duxford from the Isle of Wight, awaiting the completion of the wings and eventually the mating-up of wings and fuselage. On 13 August 2008 the aircraft was again re-registered, this time to Thomas Kaplan's Mark One Partners LLC, the ultimate name of the company that had been set up to own and commission the build of P9374 and her eventual hangar mates. The Spitfire had gone from loss, to disappearance and into obscurity, and from there to eventual discovery to recovery and thence back to relative obscurity again. Now, she had been well and truly re-discovered.

As regards the civil registration of P9374, G-MKIA, this is the *official* identity of the Spitfire and its registered designation under the UK's civil aircraft register. Whilst we will look in the following chapters at the complex path to re-creating this aeroplane, it is important to note that cosmetically and for all other outward appearances she is P9374. Special dispensation was granted for this aircraft (as well as for other restored and preserved historic 'warbirds') to wear military service RAF warpaint in an original scheme and colours, and for it to bear no other mark than its former military serial number P9374 and the fuselage code J. Let us now look in more detail at how the pile of wreckage hauled out of a French beach finally became that aeroplane again.

8 THE SEARCH

The Spitfire was certainly the most famous fighter aircraft of World War Two and quite possibly the most famous aircraft of all time. It was in continuous production for over ten years, and it remained in operational service with the RAF for nearly twenty years. Given this record, and the numbers built and the once vast holdings of Spitfire spares stock one might think that the acquisition of genuine Spitfire items, even at this distance in time, might be relatively simple. If only that were so!

Not only has the Spitfire been out of production and out of service for long over half a century, but any spare parts that might still exist are searched for and sought after on a worldwide basis by an ever increasing army of collectors, enthusiasts, museums, aircraft operators and restorers. This evergrowing band is seeking an ever-dwindling supply of Spitfire material; material which is consequently increasing in value with what is perhaps an unsurprising rapidity. Little wonder, then, that when some refer to Spitfire aircraft on internet pages they do so using the descriptive $pitfire$ – an acknowledgement of the value not only of Spitfire parts but of Spitfires themselves. Currently, a ball-park value for a 'standard' Spitfire is something around the mark of two million pounds (Sterling).

In addition, at the other end of the spectrum, Spitfire collectibles continue to command high prices, too. In fact, and as this book was being written, a well-known auction house was offering for sale as a single lot in a forthcoming auction a complete camshaft assembly said to have been taken from Spitfire P9374. The 'provenance' with this item stated that it had been retrieved in June 1971 by two enthusiasts who had travelled across the English Channel in a 1943 DUKW amphibious vehicle and removed the camshaft from the engine.

In fact, the camshaft could not have originated from the engine of P9374 and there were several reasons for this. First, the wreck was buried under the sand in 1971 and did not appear until 1980. Not only that, but *both* camshafts were still present in 1980, when the wreck was recovered in 1981 and when it later went into storage. The episode, however, shows the continuing hunger for Spitfire relics and had the item been sold with this provenance (it was later withdrawn from the sale) it is likely that it would have realised several hundred pounds simply by virtue of its association with P9374, a now world-famous Spitfire.

Notwithstanding the fact that 20,351 Spitfires and 2,408 Seafires were built, the scarcity

A period photograph showing the cockpit fit for an early Spitfire I.

of parts is perhaps not surprising in the early part of the twenty-first century. After all, the only value in these aircraft postwar, or post-service, was for scrap. Very few were ever earmarked for museum preservation, and the rest were simply broken up for scrap metal. The same fate befell literally thousands of tons of spares, although thankfully a great deal of equipment and instrumentation etc. was sold off as government surplus rather than as scrap. There were, perhaps, two reasons for this.

First, there was little scrap value in what were mostly Bakelite and glass instruments. Secondly, there was a continuing market for such items in the postwar aviation world, as a great many aircraft flown in that period were either ex-military types (for example, Tiger Moths, Avro Yorks or Lancastrians and DC-3 Dakotas) and thus continued to use exactly the same wartime instrumentation that had been fitted in RAF aircraft. Fortunately, in-strumentation across many wartime RAF types was standardised; thus, many of the same instruments could be found in Spitfires as well as in a wide range of other RAF aircraft. Consequently, the stocks of such items that were retained postwar have mostly resulted in there being a relatively plentiful supply of flying and engine instruments still being available. That said, there are exceptions and there are difficulties when it comes to sourc-ing some instruments for Spitfires and it is the instrumentation and cockpit fit that we will look at first.

In the case of the P9374 project those difficulties were compounded by the fact that this was an early Mk I Spitfire and the restoration required an exact stock-standard fit of instrumentation. In other words, and even if the instrument type and model were exactly as per the fit that P9374 would have had, those instruments were rejected if, for example, they happened to carry a date such as late 1940 or 1941. Such was the attention to detail! So, not only was there a challenge to find the exact type of instrument but it must also carry the right date even although, in the majority of cases, those dates would not be easily visible or else would be completely hidden. Moreover, to manufacture brand new instru-ments would be both challenging and hugely expensive and not generally an option. How-ever, there were one or two instances where cockpit-fit items had to be modified if not

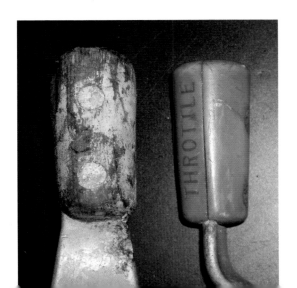

No clear evidence existed as to how the throttle lever on P9374 should be finished, as examples from early mark Spitfires existed in two patterns and fin-ishes; white painted wood or engraved bone, as seen here.

re-manufactured. First, though, it was necessary to establish exactly what the cockpit fit of a Spitfire I, direct from the late 1939 production line, might have been. And that was a surprisingly difficult task.

Whilst illustrations and details in the pilot's notes and other RAF air publications exist in profusion, there is a considerable lack of information appertaining to much of the precise and intricate detail of the cockpit. For example, the little wooden handles on the throttle and propeller pitch control show up as being painted white in an early instructional film of the Spitfire I cockpit. However, the evidence as to the finish on this particular control lever is far from clear insofar as what might have been the case in P9374. Certainly, white painted levers are extant on some surviving throttles whilst others exist that are made of a bone-type material and with the word 'Throttle' engraved onto them. In fact, former 92 Squadron Spitfire pilot Geoffrey Wellum was absolutely clear in his memory of the bone-handled variety, although he did not recall ever seeing the white painted wooden type. In the end, however, and notwithstanding Geoffrey's recall, the reconstruction team elected to go with the white painted wooden version as it has subsequently become clear that, variously, both patterns were used on Spitfire I aircraft. Although this seems a minor matter, it was attention to detail such as this that would recreate P9374 in her required stock-standard 1940 configuration. In this particular instance, however, no absolute conclusion could be reached as to the correct path to take and this

Enthusiast Steve Rickards diligently and painstakingly researched the accurate cockpit fit for P9374, thus enabling the Spitfire to be equipped just as it was on 24 May 1940. Here, Steve poses in original 1940 RAF flying kit in front of the restored P9374 at Duxford and thereby adds some period 'flavour' to the scene. He is wearing exactly what Peter Cazenove would have worn when he landed on the Calais beach.

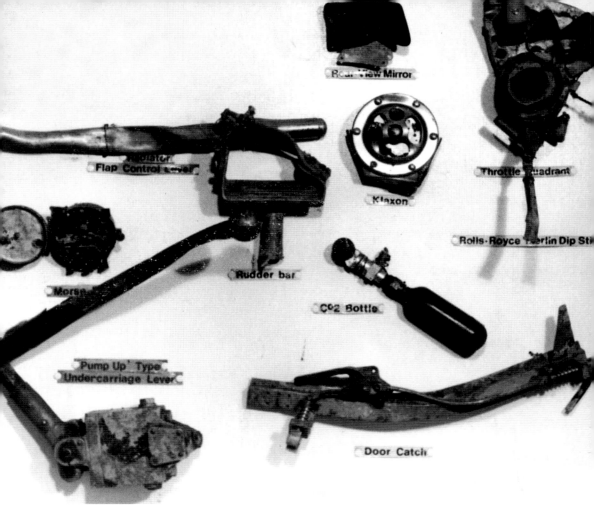

Rear View Mirror

Flap Control Lever

Throttle Quadrant

Klaxon

Rolls-Royce Merlin Dip Stick

Rudder bar

Morse

CO2 Bottle

'Pump Up' Type Undercarriage Lever

Door Catch

Wreck-recovered items from Battle of Britain Spitfire crashes suddenly became a vital resource in working out what would have been fitted in P9374, as well as being the source of some original and otherwise hard-to-find items that were ultimately used in the reconstruction process.

is illustrative of the many puzzles and challenges that needed to be resolved. So, in order to get it just right, the project engaged the services of enthusiast Steve Rickards who conducted what can only be described as a painfully detailed study of the tiniest aspects of Spitfire I cockpits.

Part of the problem, even with the very few extant museum examples of Spitfire Is, was that their cockpit fits were no longer as they had been on Spitfire Is straight from the production line. The reason was simple, and was linked directly to the fact that in-service all RAF aircraft are subjected to regular modifications that would be applied, as it were, to the machines in the field or else (in the case of more major modifications) at maintenance units or factories when the aircraft were returned for repair or overhaul. This meant that a series of upgrades throughout the life of a Spitfire could result in fairly significant changes to an airframe that made it little resemble, in many respects, how it looked as a

Relic Spitfire I instrument panels like this one proved invaluable in Steve Rickards' quest to establish exactly how the panel on P9374 was fitted. In fact, such items were just about the only available and reliable source for this information!

brand new machine. A case in point, especially with P9374, was the fitment of its hand-pumped undercarriage handle. Had she survived beyond May 1940 then this Spitfire would have been fitted with a powered hydraulic selector. And, as we have seen, P9374 had to be exactly as she was when she hit the sand at Calais.

Consequently, Rickards needed to turn to original and genuine 1940 material for his primary research tool and, aside from photographic evidence, he did so using wreck-recovered material that existed in various private collections and museums in the UK. Suddenly, objects that many had previously regarded as just having the intrinsic significance of historical artefacts now took on a whole new importance. They were just about the only surviving hard evidence that told us exactly what a 1940 Spitfire I cockpit might have looked like, and Rickards studied every single shred of such evidence that he could lay his hands on with an astonishing forensic exactitude.

Steadily building up a picture of a 1940 cockpit fit, it didn't take him very long to realise that there were even anomalies between how some Spitfire Is looked when delivered, brand new, to the RAF. In fact, the instrument panels of some Spitfire Is seemed to differ markedly in detail to others which had been delivered from the same factory. For instance the absence or presence of either an ammeter or flap position indicator were variations between various Spitfire Is. If decisions had been taken (for whatever reason) by Vickers Armstrong during the production run of Spitfire I batches to modify certain items of construction, systems or fitting details, it quickly became obvious that no detailed surviving record existed of those changes, albeit that some differences could be linked to known modifications. Somehow, such minutiae had to be deduced through detective work – and sometimes just a best guess!

This was the cockpit of P9374 immediately post-recovery. Everything above the compass and compass tray had disappeared. The broken-off control column top can be seen left centre.

The problem for Steve Rickards was, of course, that the recovered wreck of P9374 had no instrument panel above the level of the bottom of the panel, there being nothing remaining above the compass tray. It was literally a blank canvas, and a little bit like a painting-by-numbers kit but without knowing exactly what colour went into which space. Of course, the general layout was known and well recorded, but even such detail as the actual *finish* of the instrument panel itself had to be deduced, along with the actual process of that finish and how the written instructions were applied. Were they etched? Or had they been engraved? It all had to be worked out, and then it had to be turned into reality.

Measuring, photographing and drawing Spitfire I instrument panels (or fragments thereof) consumed many hours of Rickards' time, but ultimately he came up with drawings, charts and spread sheets that enabled Martin Overall at Historic Flying to produce and fit out an instrument panel that mirrored, as exactly as humanly possible, just how

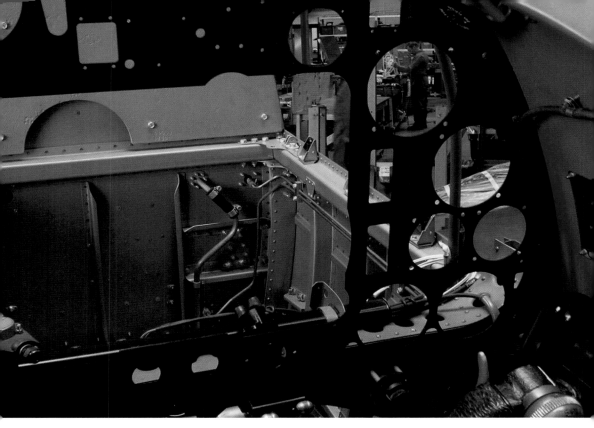

The new-fit instrument panel in place at Duxford and waiting to be kitted out with its instruments and controls.

P9374's had looked in May 1940.

Whilst some of the instruments were relatively easy to source they still had to be over-hauled and calibrated and declared airworthy. Others, though, were more of a challenge and items that are always difficult to find are Spitfire undercarriage position indicators and the nose up/nose down trim gauges. Both of these instruments were unique to Supermarine aircraft, and thus produced in far fewer numbers than the 'generic' instrumentation that was common in a much wider range of RAF types. In the case of the trim gauges, for example, it would be fair to say that these are as rare as proverbial hen's teeth and the undercarriage position indicator, whilst still difficult to find, is slightly easier to obtain although now commands many hundreds of pounds or dollars to purchase. So much for the instrument panel; what about the other cockpit fittings and equipment in this aircraft? And given their rarity where are all of these bits and pieces to be sourced in 2010 and 2011 if items original to P9374 cannot be used?

Perhaps the most visible item in the Spitfire cockpit is the very prominent Barr & Stroud reflector gunsight, positioned right at the top of the instrument panel directly in front of the pilot and sitting just behind the armoured windscreen. Its functionality when fitted in a Spitfire flying during the early part of the twenty-first century is superfluous and entirely cosmetic, although the exacting brief to produce P9374 as a 'stock' 1940 aero-

Above: The fully fitted out instrument panel with its complement of original 1940 dials and guages.
Below: An original reflector gunsight was obtained and fitted to P9374 and is in fully working order.

plane meant that one would need to be sourced and fitted.

Unlike with instruments, of course, there was no particular postwar use or purpose for aircraft reflector gunsights although by great good fortune an Essex-based optical com-

pany, Messrs C W English & Co, had purchased large numbers of these items as government surplus, and still in their wooden Air Ministry transit boxes, when the company recognised that the sights had a commercial value because of the various high quality lenses contained inside them. Messrs English & Co still had large quantities of unused and still-boxed sights during the 1970s and they began to sell them to the enthusiast market and also in bulk to militaria dealers for £25.00 each. (Now, these same sights command prices of several hundreds of pounds.) Whilst all of English's sights had long since been sold off by 2011, there were sufficient examples in widespread circulation that gave the team hope that sourcing one would not prove too much of a hurdle. Of course, it didn't turn out to be so.

One of the problems with the gunsights from English's warehouse was that they were mostly of a slightly later pattern than was specifically required for P9374, and pretty much all of them had a square reflector glass rather than the oval screen that was fitted to the earlier Spitfires. That said, in the 1970s it was a case of paying your money and taking pot luck with what Messrs English supplied; sometimes, although very rarely, the buyer got lucky and received the sought-after and coveted type which had been used in 1940 and of exactly the same type as on P9374. To be precise, the Reflector Sight Mk II, stores reference 7B/1124. For P9374 nothing else would do, and by great good fortune one was located with Hurricane builder Tony Ditheridge who was able to supply this treasure for the project. Still in its transit box, still pristine…and *still* working! Fitted in place in the cockpit of P9374 during 2011, the sight switches on and illuminates, brightly throwing up the orange light ring with its cross wires onto the reflector screen – that feature alone transporting back to 1940 those who have been fortunate enough to peer through the glass. Of course, the light is reliant upon special 12-volt bulbs and quite remarkably a source of new-old stock (hitherto unused contemporary material) was located, not only enabling replacements to be obtained for the future but also allowing the standard set of spares to be mounted in a clip on the cockpit wall.

Second only to the gunsight in terms of the visible and iconic items in the cockpit is, of course, the control column – most notably its spade grip top and the gunfiring button with its SAFE/FIRE ring. Yet again, these items are extremely rare. Those few that do exist are either in museums or held in private collections; jealously guarded by a few and wantonly coveted by many. When originals do turn up on the market they command many thousands of pounds and are gone almost before they are offered for sale. Again, there are two main elements that account for this rarity when set against control column tops, say, that would have been fitted to later mark Spitfires. Primarily, it is because far fewer were ever produced than were manufactured for the more prolific later types. Not only that, but they also differ quite significantly.

The early Spitfire (and up to the Spitfire Mk Va) spade grip tops had a round brass gun-firing button, encircled by a rotatable safety ring in a red anodised alloy and marked SAFE or FIRE. Later marks of Spitfire, from the Vb onwards, had an oblong alloy 'rocker' button and the outward appearance was really quite different to the component that would have been fitted to P9374. Yet again, the original control column top for P9374 no longer existed and was gone when the author inspected the wreck in 1980. In fact, there was fairly clear evidence that it had long ago been 'souvenired' (probably by the Germans) but even if it had unusually been left in situ by the curious trophy hunters of 1940 it could not have survived until 1980, anyway. The reason being simply that the metallic composition, under the vulcanised and patented Dunlop rubber coating, was of a magnesium alloy material.

Whilst light, functional and durable in 1940, it is a metal that does not last well (if at all) when buried in the ground or submerged underwater and it dissolves over years.

For the same reasons, therefore, no intact examples have ever been recovered from the many excavated wartime wrecks of Spitfire I aircraft. Electrolytic processes have conspired to make very much rarer the Spitfire I spade grip, and all that aviation archaeologists have ever found on countless 'digs' has been crumbled white powder that had once been the spade grip and with just the brass air lines, aluminium brake lever, firing button and alloy safety ring surviving intact. A complete grip has never been found on recovery excavations, unlike in Hurricane wrecks where the almost identical-looking spade grip frequently survives in near mint condition, simply on account of the fact that they were manufactured out of a cast aluminium alloy rather than a magnesium alloy.

This mirrors the challenges which time and again faced the P9374 re-build team. However, when no original spade grip could be sourced in time for the completion and first flight of P9374, a composite look-alike substitute had to be constructed using a similar gun button and a similar pattern grip. To date, however, P9374 awaits an original type AH2174 spade grip, and although one will doubtless be found in the future it must, of course, be to a fully airworthy standard and in tip-top condition.

Whilst sourcing a spade grip from an archaeological dig or recovery proved a nonstarter, it is not the case with other artefacts and items that have proved useful in the build of P9374. Instrumentation aside, there are countless levers, fittings and information plaques within the cockpit that have been found deeply buried in the ground and are still in a usable and airworthy condition given careful examination and usually minor attention. A case in point might be the flap or landing light levers which are small chrome-finished levers looking somewhat like old fashioned gas taps. Generally, these items survive well in crashes but are otherwise difficult to source. Again, they are Spitfire-specific and thus on the ultra-rare list of must-have Spitfire hardware. Equally Spitfire specific are the rudder pedals, grandly embossed in the forged facing of the alloy pedal with the inscription 'Supermarine'. In the case of P9374, both pedals were extant when the author

One of the original rudder pedals may just be seen almost dead centre in this shot of the cockpit of P9374 at Le Bourget.

got into the cockpit at Calais in 1980 and both were still there, apparently, when the wreckage arrived back in the UK. Sadly, one of the originals has since been stolen and has had to be replaced with a copy. A great shame, indeed.

Embossed with the stylised logo, items like these are attractive trophies for enthusiasts. This particular pedal was recovered from a Spitfire shot down over Romney Marsh during the summer of 1940 and shows to advantage the detail of this important item. The canvas strap was simply folded-over fire hose!

So, apart from museums, collectors, other restorers and aviation archaeology where else did the P9374 team discover all that it needed to complete an authentic 1940 Spitfire? Sources often include the now rather fewer 'aerojumble' events, where relics and ephemera can be found in relative abundance. Even now, useful and usable Spitfire parts can turn up at sales like the regular Shoreham Aerojumble, but long gone are the days when there was almost a surfeit of such material at sales like the Yeovilton Aerojumble of the 1980s and where it was sometimes difficult to give such items away! Apart from such sources, there are dedicated historic military aircraft parts traders on a worldwide basis (including one dealing predominantly in Spitfire spares) and those who built P9374 scoured all and every last known source, and more than just the once.

As has been stated, the owners of P9374 had given a very exacting brief in terms of what should or could be used on the aeroplane, and essentially this required the builders to use as many parts as possible from the original recovery as provenance and originality (where possible) was paramount. If original parts could not be used from the actual aircraft then, as second choice, they should be sourced from a similar airframe or another spares source. Making a new part was the final fall-back position, unless, of course, modern safety provisions took priority over original fitment. An example of the latter would be the fitting of a full CAA safety-approved modern pilot's seat harness and the deletion (for obvious reasons) of the wartime canvas Sutton seat harness, although an original harness is retained by the owners of P9374.

In terms of sourcing original parts, however, there is no better example than the armoured windscreen – or, rather, its frame. When the author inspected the wreck at Calais in 1980 he was given the full armoured windscreen, complete with its coaming and armour-plated bottom strip. Clearly, this was an item that *had* to go back to P9374 where it was of inestimably more use and value than sitting in the author's study. Refurbished, the frame seen today on P9374 is exactly original although, of course, the armoured glass had to be replaced

A pile of newly re-made tyres from Dunlop as they were delivered to Duxford.

with new as the original was cracked and crazed and opaque through decades of saltwater and sand immersion. The coaming, too pitted to use, was nonetheless a valuable pattern for the fabrication of a replacement and as with all the original parts that were not incorporated in the reconstruction, it was still retained by the aeroplane's owners. Nothing was thrown away. After all, these unused but original parts are the very essence of P9374's history. As to re-manufactured items, a notable example must be the main landing gear tyres.

Whilst an option clearly existed to use look-alike substitute tyres, this would not do for the owners. Nothing but original balloon-type smooth Dunlop tyres would suffice and although the original tyres were recovered with the wreck in surprisingly good condition they were not serviceable. Of course, even if new-old stock tyres of the correct specification still existed in some obscure warehouse store, they would be wholly useless for an airworthy Spitfire today. Perishing, cracking and general degradation of the rubber would have seen to that. In any event, tyres are 'consumables' and a ready supply of replacements needed to be sourced. Thus, the only option was to have a large supply of tyres made to the original specification (at huge cost) and marked with the distinctive 'Dunlop' logo, picked out in white. Here, then, are three examples; originality sacrificed to comply with twenty-first century safety, the re-use of a part wholly original to P9374 and the re-manufacture of an otherwise unavailable item.

Surprisingly, quite a few parts that could otherwise not be sourced have turned up on internet auction sites and quite apart from having to compete against others who keenly want whatever item might be being sold, it is often the case that a vendor has no idea that

In place on the main landing wheels, the tyres are resplendent with the Dunlop trade name distinctively picked out in white.

they might be selling a Spitfire part. Consequently, such items might have been wrongly listed and not widely seen or noticed by other Spitfire hunters or else sold way below market value. Such instances are now quite rare in this internet-enlightened age, and it is more likely the case that a vendor will know exactly what he has and trumpet the magic word 'Spitfire' in his sales pitch, thus ensuring a feeding frenzy of buyers and the highest possible prices. In fact, more often than not a vendor will claim Spitfire provenance even if the item is from something like a Cessna 172 or De Havilland Vampire. On the other hand, optimistically type 'Spitfire wing' into an internet search engine and one will, of course, be directed to a good many wings for Spitfires, but they will not be port or starboard 'wings'. Instead, expect to be offered nearside or offside wings for *Triumph* Spitfires! Such are the trials and tribulations of the Spitfire hunter.

The detective work required to achieve the splendid result that is P9374 today has not only been embodied in the physical hunt for genuine and period Spitfire components, but also in a journey of discovery looking for the detail of exactly and precisely how a Spitfire I appeared in May 1940. It was not an easy search.

Not by any means had all the required fixtures and fittings for a Spitfire I been sourced by the time that the construction of the fuselage and wings commenced. However, this was viewed as 'work in progress' that could proceed in tandem with the comparatively easier task of building the airframe. Whilst production lines and the various jigs and formers to build Spitfires in their original factory settings had long since gone to the scrap yard, the process of Spitfire construction in the early twenty-first century is now a well tried and well understood journey. Notable in the field of Spitfire reconstruction and restoration expertise is Steve Vizard's Airframe Assemblies on the Isle of Wight, and it was this company that won the contract to build the fuselage for P9374 – and only just a few miles across The Solent from where this aeroplane was first constructed.

As a specialist company in the field of historic aircraft construction, Airframe Assem-

Original Spitfire I construction with new fuselage assemblies in their jigs circa 1939.

Similarly, new Spitfire I tails are shown here in their construction jigs during the 1939/40 period.

blies holds UK Civil Aviation Authority approval to British civil aviation airworthiness requirements (BCAR) A8-2.A2. This makes the company almost uniquely well placed and with the authority, where appropriate, to release components and sub-assemblies to that approval standard. Full release is not and will never be possible for old warbirds because there are no longer any design approval holders (ie the original designers and builders) left officially to sanction design changes using modern materials etc. Procedures within the company are geared to a uniform level to meet the BCAR A8-2.A2 standard and this enables Airframe Assemblies to supply its own certificates of quality assurance based on those standards. Thus, it will be seen that there is a very great deal more to reconstructing an aeroplane like P9374 than might at first be supposed.

The building of Mk V, IX and numerous other Spitfire variants was almost run-of-the-mill (notwithstanding the exacting standards set out above) and had pretty much become bread-and-butter jobs for Steve's firm, but there was an interesting element thrown into the contract to build a Spitfire I; namely, it hadn't been done before. At least, not since 1940 on the Vickers-Supermarine production line. Although all of the basic skill and knowledge was there within Airframe Assemblies to build a Spitfire fuselage, there were details that varied quite considerably on a Spitfire I from, say, a Mk V. As an example, the IFF aerial lead run-in ports on either side of the fuselage that took the aerial wire from the tip of each tail plane into the fuselage. Standard, almost, on many other Spitfires they simply weren't present on a Spitfire I and inadvertently were initially incorporated into the build. Of course, this had to be rectified although it was hardly a clumsy mistake and neither was it a major one. It was, though, typical of many 'corrections' that had to be made along the way in order to get things 100% accurate.

One key to achieving that accuracy was not only Steve's extensive knowledge of the Spitfire but also his *intimate* knowledge of early Spitfire I and II construction. This was an intimacy gained through many years of excavating and collecting Battle of Britain aircraft wrecks, something which had been an almost obsessive passion for him during the 1970s and 80s. That interest, by its very nature, had given Steve a great deal of awareness about how early Spitfires were constructed what even the tiniest component was, and even what particular shade of a particular colour they might have been. When it came to knowing that something looked right, or that it didn't, one could not have hoped for a more efficacious apprenticeship in Spitfire I structures. And there wasn't very much left of the P9374 fuselage structure, at least not of any significance, to use for templates or patterns.

Of course, a relatively extensive set of detailed engineering drawings and other general arrangement drawings for the Spitfire I still exists, as do parts lists and other RAF air publications which are regularly drawn upon as the only available documentary, graphic or measurement and dimensions source. That said, the value of relic or unusable items can often be incalculable, not only in seeing exactly what a particular component looked like but, sometimes, how it was put together. All of this, of course, is inevitably painting a picture that might make the reader puzzle over the 'originality' of the eventually reconstructed P9374. It is always an area of contention within the historic 'warbird' scene, and amongst the thousands of followers and enthusiasts of such aeroplanes who cherish these old aircraft there is a frequently aired question: what is 'original'? It is a question that cannot be ducked or avoided within the context of this book and the story of P9374.

However appealing might be the notion that Spitfires, Hurricanes, Mustangs, Lancasters and the like that are seen wheeling and soaring at air displays are entirely original, such is not the case. It cannot be. Take, for example, the RAF's renowned Battle of Britain Memorial Flight. Whilst the majority of those aircraft have been on the RAF's books continuously since seeing operational service, it is difficult to describe them as wholly original, or even anything like it. In getting on for the passage of seventy years, each aircraft will have been re-skinned, re-rivetted, re-sparred, re-fabriced, re-engined and just about re-built a good many times over. Had they not been they would long ago have been relegated to static museum pieces.

A particular case in point is the memorial flight's Hurricane, LF363, which suffered a disastrous engine failure, crash landing and subsequent fire that all but destroyed this valuable aeroplane. Despite its charred and mangled condition, the Hurricane was fully rebuilt and restored and is now flying again with the flight. Sometimes, the popular analogy of 'Trigger's broom' is used to draw a comparison; his broom having had so many new heads and handles that describing it, as he did, as the *same* broom might be called into question. That said, and if Trigger had had an extensive enough vocabulary to express such a view, then he might have said that it was the same broom because it had a 'continuing provenance'. Perhaps that is a rather extreme example, but the reader will surely understand the gist. And it is much the same with the aircraft of the Battle of Britain Memorial Flight; all have a continuing provenance. And certainly (at least in most cases) all of them have retained a good deal of their original equipment, fixtures, components and fittings. That provenance, and a continuing one, is the all-important factor.

In the case of P9374 the restorers were faced with a wreck that had been positively identified as that aeroplane, and although the ultimately recovered wreckage was very far from a 'restorable' Spitfire *per se*, here were the building-blocks (quite literally) for a newly reconstructed Spitfire that continued the original identity through the inclusion of significant parts original to that unique airframe. In any event, and had P9374 still looked exactly like a Spitfire when she had been delivered to HFL at Duxford, the end result of the reconstruction process would have been exactly the same.

In other words, it hardly mattered that P9374 was now just a jumble of bits, because even had that jumble still held the appearance of the original machine it would have made no difference to what would need to be done or what could still be saved and used to enable it to fly again. Quite clearly, none of the original skinning structure could be used again. It hardly needs an explanation as to why. Hence, the outer shell of P9374 is, by definition, new. It can be no other way. In fact, had the relatively pristine Spitfire I (R6915) been lifted down from where it is on display in the rafters of the Imperial War Museum at South Lambeth in order to be restored to flight there can be little doubting that the process to return it to the sky would very much mirror that for P9374. True, there would probably have been a very much greater stock of original and usable parts but the engine, propeller and pretty much all of the airframe structure would, essentially, have to be built as new in this hypothetical project. This, however, raises an interesting issue in relation to the originality of surviving Spitfires like reconstructed aeroplanes such as P9374. And Steve Vizard has an interesting point of view on that. Even in wartime, there would have been hardly any original (that is ex-factory) Spitfires left by the end. Let's follow his scenario through.

If ones takes, say, a hypothetical in-service Spitfire V, it is quite likely that on delivery to a maintenance unit it might have been noted that a pre-service issue modification had not been carried out and so this would be rectified. On delivery to its squadron, perhaps a series of other mods would be conducted before combat damage resulted in a new tail section being fitted along with some new fuselage and wing skins. Later, a new engine fit could result in the original cowlings being misplaced and new ones being fitted. Back in service, and if damaged, a tail wheel oleo leg would be replaced with new, whilst a major overhaul would see many other upgrades and mods before the Spitfire was transferred to an operational training unit. Here, it would almost inevitably be 'pranged' by a trainee pilot and extensive although repairable damage caused to the under wing scoops, cowlings and centre section. The result would be an extensive rebuild by a civilian repair organisation facility, to

WARNING
DO NOT MOVE
CHASSIS LOCK
LEVER WHILE
AIRCRAFT IS
ON GROUND.

The hand-pumped hydraulic undercarriage lever on the starboard side of the cockpit, unique to early Spitfire I aircraft.

the extent that when the three-year-old Spitfire finally went back into service it would be very far from the original Spitfire that had left the factory. So, as Steve postulates: "Was it still the original Spitfire?" The answer has to be no. Indeed, in his view, many restored or reconstructed Spitfires are, in many respects, more original than the originals!

So, whilst the fuselage for P9374 was built incorporating as many parts of the type original to 1940 as possible, as we have seen, Steve takes the view that the end result is perhaps more original than an in-service aeroplane of the period. Certainly, more original in fit and finish than surviving museum Spitfire Is of the period.

As we will read in more detail later on, Steve was absolutely correct in saying that all service aircraft were (and are) subject to modifications during their lifetime. These were changes brought about by problems that had perhaps come to light through operational experience. For example, a particular item may have been found to be weak or faulty and would be replaced, in the field, by a newly modified or strengthened item.

Similarly, an improved or updated version of a particular piece of kit might be adopted for newly constructed Spitfires and, as a consequence, then be incorporated as a mod to existing airframes. A case in point, and a very significant one, is the change from a hand-pumped undercarriage selector to an engine-powered hydraulic unit. Those Spitfire Is in service that had the pump-up type were eventually modified to be fitted with the powered type, and that modification alone resulted in the re-positioning of some other items on the starboard side of the cockpit such as the reflector gunsight spare bulb stowage clip and the Morse code tapper – yet another modification! So a Spitfire I that had survived

1940 and was still in service by 1941 even if it had not crashed or suffered damage, would still have significant differences compared to how it had left the factory and would have had many dozens of modifications. An example of this is the 'restoration' in recent years of the RAF Museum's Spitfire I, K9942, by Lewis Deal's Medway Aircraft Restoration Group on an aeroplane that the restorers retrofitted with its pump-up undercarriage lever which had in service been modified with an automatic selector. Here was a classic case of a Spitfire I, long in operational service, and 'modded' to the ultimate degree. To give but few other examples, some of the modifications that would have been applied to any Spitfire I that had survived beyond May 1940 would have been: Mod 247 – armoured glycol tank (6 June 1940), Mod 283 – Fit Spitfire Mk III windscreen and hood (27 July 1940), Mod 422 – Introduce ¼ inch droop metal covered ailerons. These, then, are but very few of many dozens of modifications that would have been applied to a Spitfire I, and had an airframe survived the war then those changes would have been legion by the end of its service life. Thus, these alterations would make the use of any surviving museum Spitfire I to be of limited value as a study piece or pattern.

Conversely, a few modifications were effected pre-May 1940 and an example of this was the deletion of the flap position indicator (Modification 215) which was ordered on 20 March 1940. This explains the empty round aperture in the top left-hand corner of the re-constructed P9374 instrument panel, clearly where an instrument *should* have been fitted. This empty hole might well give the appearance, to the uninformed at least, that somehow an instrument has been missed out! So, the challenge was as far as possible to put it back to its *original* state, and Steve Vizard's point was that the reconstructed fuselage of P9374 would be stock standard for a factory-fresh aeroplane and, of course, modification free, unless they were mods put into place before P9374's demise. When it came to attention to detail, though, much of that was manifested in the fitting out and painting of the Spitfire, as well as the attachment of various items of equipment.

The fitting out of systems etc. was a task undertaken at Duxford, as was the attachment of data plates and instruction plaques. Working out what some of them said, let alone exactly where they should be fitted, was another challenge to be faced and, unfortunately, the majority of such plates had vanished long before P9374 came up

Construction data was found still stencilled onto part of the recovered wing flap structure of P9374.

out of the beach. However, the author had noted the details of the very few data plates present when he viewed the wreck in 1981. These were as follows:

Elevator: 6/1/40

30020   Serial No: AE 6S 48891

Mod Plate

Serial No: GAL/6S/58300

30064/325

Singer Motors Ltd

25/1/40   Serial No: 59625

AID CU2

The Heston Aircraft Company Ltd

HA 6S 44269   20/1/40

R/H Inner

The above were all sub-assembly construction plates, the one above right being from one of the wing flaps. Whilst they were just tiny details they would ultimately all help in the replication of accurately represented parts for P9374.

Perhaps, more than any other project, P9374 has amply demonstrated the value of 'aviation archaeology'. An activity that has often been questioned in terms of its purpose and value, it is an absolute matter of fact that were it not for aeronautical wreck recovery P9374 would have never become the successful project that it did. Moreover the quest of people like Steve Rickards, and others involved in the build and fitting-out at Duxford, would have been completely impossible were it not for the number of recovered Spitfire I wreckages that were available for inspection. Quite simply, without that study material it would have been impossible to determine many of the all-important and intricate details. What exactly did this bit look like? What colour was it? How, and where, did it fit? Was a Spitfire I fitted with one of those? And what size, shape or profile was that bit? Manuals and drawings gave lots of good clues and pointers. Photographs helped. Surviving Spitfire Is (such as they were) gave some assistance. However, it is no exaggeration to say that in this project the examination of wreck-recovered material was vital. As we will see in Chapter Eleven this was absolutely the case, too,

The colossal schedule of spare parts for the Spitfire I listing all of its thousands of individual numbered components. With it is one of the RAF Spitfire I Air Publication manuals. These rare documents were an invaluable resource in the reconstruction of P9374.

HAND TURNING GEAR
FOR MAINTENANCE ONLY

IF USED FOR EMERGENCY STARTING,
AIRCRAFTSMAN MUST HAVE ROPE FROM
HIS WAIST TO THE UNDERCARRIAGE
TO PREVENT HIM FALLING INTO THE
AIRSCREW.

with the airscrew re-construction. Very appropriate, then, that Steve Vizard's roots (pre his establishment of Airframe Assemblies) had been in aviation archaeology where he had amassed a good deal of extremely specialised knowledge of the Spitfire I, not to mention a goodly collection of period parts.

Valuable parts were also in the author's collection and played an important role in getting P9374 'right'. Take, for example, an oblong instruction plate fitted on the starboard engine cowling just below the hand-cranking aperture. In photographs of early mark Spitfires this is a clearly visible and light-coloured plaque with seven lines of indistinguishable lettering. The question was; what did it say? Nobody seemed to know. The puzzle was finally solved by the author who remembered he had found one during an excavation to recover a Spitfire that had been carried out in Sussex during the 1980s. He still had it, and the wording on this small item was quite bizarre and extraordinary.

The author's example (too poor to use on P9374) was adopted as a pattern in order to draw and then re-manufacture a new one, and this was a task taken on by Guy Black's historic aircraft workshops at Retrotec in East Sussex. This tiny attention to detail was entirely typical of the work carried out to make P9374 precisely accurate in every respect, and whilst some wartime Spitfires later than the Mk I

Above: The instruction plate (top) for hand-turning the Rolls-Royce Merlin engine was fitted to the starboard engine cowling. An original pattern was supplied by the author, thus enabling exact replications to be made.
Below: This relic gun-bay warning plate was another original 1940 item used as a pattern to re-manufacture exact replications for P9374.

have incorporated this instruction plate, none of those currently restored and flying have been fitted with this tiny item. In fact, it is clear that some other restored Spitfires have had new cowlings manufactured and the redundant hand-turning aperture is then simply deleted – along with the adjacent instruction plaque. Not so for the P9374 reconstruction.

Similarly, other data plates like those in the wing gun bays could not be sourced as originals and those that had been original to P9374 had long since gone. Once again, the detail of what they looked like and their inscription could only be determined by using a wreck-recovered item supplied by the author as a pattern and thus allowing exact copies

Although photographed in 1941, and with a slightly different paint scheme to P9374, this unique colour photograph is a good representation of a period Spitfire, albeit that this is a Spitfire II. The aircraft is from 54 Squadron and interestingly is pictured on 24 May 1941 at RAF Hornchurch – exactly one year to the day after P9374 left the same airfield. The small black dot in the white of the fuselage roundel, just to the right of the airman, is significant (see page 90).

to be made. Such was the detailed work carried out on the re-manufactured parts that even the unique part number, 30008 47330 21263, had been stamped almost invisibly into the re-made plate, and exactly as per the original. Truly, this was attention to detail in the ultimate degree. These, then, are some of the *details* of the reconstruction. But what of the actual construction of the fuselage itself, and of the wings?

As with any restored historic warplane (nowadays colloquially referred to as a 'warbird') the structure of the airframe is, always or almost always, out of necessity an entirely new build. And as we have seen above, and quite apart from issues related to materials quality (eg, some period materials, rivets etc no longer being permitted today) it is almost certainly the case that any truly original skins, ribs, spars and coverings will have long ago passed their 'use-by' date. In the case of P9374, of course, much of the structure had been damaged in recovery. That which had been salvaged was corroded and beyond any use on a restored airframe, apart from any possible value as a pattern.

So having been awarded the contract, Steve Vizard's company set to work in 2007. Of course, wartime Spitfire production was established as a major production-line operation employing a huge workforce who turned out units in the hundreds from production-made jigs. Today, Spitfire construction is a completely different process and those Spitfires constructed seventy or more years after the originals came off the line are what might be described as 'bespoke' airframes. Each one is hand built in jigs and frames by craftsmen and engineers who lovingly create Spitfire aircraft to the client's individual specifications. Perhaps the best illustration of the construction process of the P9374 fuselage is through a series of photographs depicting the re-birth of this very special Spitfire I (see overleaf).

Whilst the fuselage had been constructed by Airframe Assemblies, the owners had elected for the wing construction to be undertaken by Historic Flying Ltd at Duxford, although, and notwithstanding that decision, Steve Vizard's operation was still called upon to supply the vital main wing spar booms. Incredibly strong, the spars were of a revolutionary design for the period and allowed for the construction of the Spitfire's almost uniquely thin wing. The square section spars comprised tubes within tubes and, from the wing root outwards, the number of tubes reduced progressively along the spar until the final tube is gradually reduced towards its tip. Currently, Airframe Assemblies is the only organisation approved to produce Spitfire spars and thus it was inevitable that they should suppy the spars for P9734; the back-bone around which the Duxford team could build the wing structure.

Sadly, the original and immensely strong 'D' box-section wing leading-edge structures from P9374, recovered in 1981, had long since vanished in France. Either they had been scrapped or been cannabilised to provide sacrificial parts for other Spitfire wing builds, but more likely the former. This was a great shame, since Martin Overall who led the

The construction of Spitfire I wings in 1939.

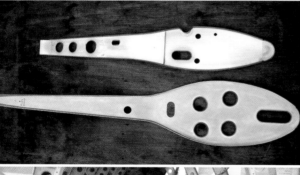

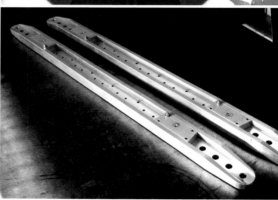

Clockwise from top left: Gradually, the skeletal structure of fuselage frames is built up to re-construct P9374; Appearing more like a fuselage as the construction gathers momentum, this is a view looking rearwards towards the fin frames; Newly machined carry-through spars for P9374. These run through the centre section of the fuselage below the firewall; Another view looking down the fuselage tube at the Airframe Assemblies workshop on the Isle of Wight. This time, it has been partly skinned; The re-constructed fin frames for P9374 at Airframe Assemblies; Like an over-sized Airfix kit, the kit of parts for Spitfire I, P9374, is laid out at Airframe Assemblies on the Isle of Wight. This is frame eleven of the fuse-lage, along with associated structures.

Clockwise from top right: The original lump of centre section structure from P9374. Surprisingly, and despite its condition, it still yielded one or two reusable parts; The tail of P9374 under construction in 2010. Again, an interesting comparison may be had with the 1939 photograph on page 77; This dramatic shot is taken looking down the completed fuselage towards the tail from the cockpit and shows all the fuselage frames in place; The fuselage of P9374 in its construction jig. It is interesting to compare this 2010 image with the photograph taken in 1939 on page 76; The fully skinned and almost finished fuselage 'tube' for P9374.

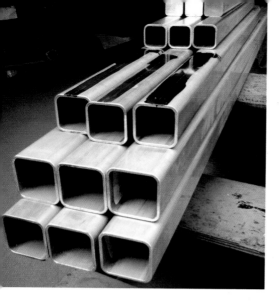

Above: The alloy extrusions which construct the wing spars. These are assembled in concentric fashion, one inside the other, to make up the top and lower wing spar booms.

Below: A cross section of the assembled tubes which makes a unique and immensely strong structure

Duxford wing-building team, was convinced that a good many usable parts could have been recovered from the originals had they still existed, but it was not to be. Instead, Martin got to work building both wings in their construction jigs along with his dedicated team including Paul 'Spike' Rivers, Martin Henocq, Chris Norfolk and Martin Parr.

As with much that was associated with the P9374 build the construction of the wings was the first time since 1940 that a Spitfire wing had been put together that had incorporated fabric-covered ailerons. In many respects, it was a retrograde step in the 21st century to fit a Spitfire with less effective ailerons as the later move to all-metal ailerons was an improvement over the fabric versions. Not only had a canvas aileron fitted Spitfire not been built since 1940, but none existed anywhere. Either flying or static.

Otherwise, the wing structure was very much the same as it was for most other Spitfires up to the Mk V (or, at least, to the Va) and broadly similar, in many respects, to many of the later Spitfire marks as well. So, there were no particular surprises or hurdles to overcome although one 'innovation' was the under-wing landing lamps in their retractable housings. Although they would never be required in P9374's present and future life they were original to the Spitfire I airframe and so they had to be there – and they had to be working; fully retractable and illuminating.

Additionally, many currently flying Spitfires are

Left: The undercarriage legs for P9374 were built up and overhauled at Duxford, and although not sourced from this particular haul of Spitfire undercarriage assemblies, it is very often the case that relic items like these can provide re-usable castings, forgings or other components. The legs shown here were recovered from a scrap yard in the 1990s and are laid out for sale at auction.

not fitted with their armaments although the brief was for P9374 to have a full fit of Browning .303 machine guns. Each wing was a four-gun wing, giving the Spitfire I its full and potent battery of eight guns. Unfortunately, and although seven of P9374's guns were recovered in 1981, they still remain in France and have not been reunited with the airframe. That, of course, is a great shame especially given that some of the recovered guns were tested and found to be in working and fireable order. For the purposes of the wing build, however, a full complement of all eight de-activated Browning machine guns has been fitted in P9374. In addition, short lengths of belted 1940-dated .303 ammunition have been connected to each gun for effect although for weight reasons the full complement of de-activated ammunition has not been installed.

Of course, when constructing the wings the all-important radiator and oil cooler need to be considered, and these were out-sourced to a specialist radiator company, Anglia Radiators. Additionally, the land-

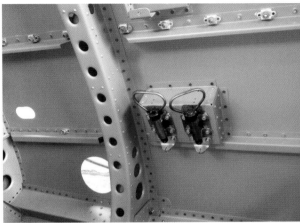

Top to bottom: The completed fuselage of P9374 is delivered to Duxford on 8 July 2008 for fitting out and eventual mating to the wings and engine; This cockpit detail shows the flare release handles which are examples of some of the original parts from P9374 used in the reconstruction project; Another original feature from P9374 is the pilot's seat armour which can be clearly seen in this photograph at Duxford. When P9374 was lost in May 1940 the distinctive triangular head armour had not yet been fitted to RAF Spitfires. The armoured windscreen frame is the original item brought back from France by the author in 1981; Martin Overall makes an adjustment in the wheel bay.

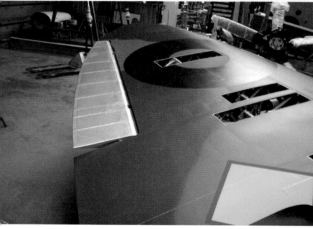

Clockwise from top: Originally, the re-constructed fuselage for P9374 had been very much to Spitfire V specifications although it was later decided that P9374 must exactly replicate a Spitfire I. Consequently, the IFF aerial lead-in socket (the black dot in the photograph on page 84) had to be removed from the fuselage which had been configured as a Spitfire II or V. This meant the replacement of fuselage panels to remove the lead-in socket, and the new panels can be seen in this photograph taken at Duxford; With the internal construction completed, the port wing is skinned in the Duxford jig; One of the wings for P9374 is reconstructed in its jig at Duxford (compare with photo on page 85); This view of the starboard wing shows to advantage the as yet unpainted fabric ailerons, unique to the early Spitfire Is. The fabric work on P9374 was carried out by Eastern Sailplanes.

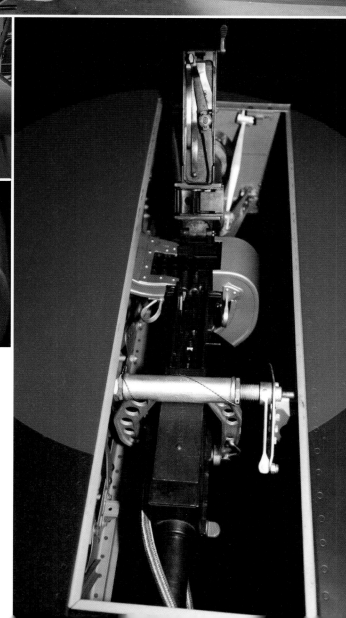

...ckwise from top: The port wing is married
...o the fuselage of P9374, although the dis-
...tive elliptical Spitfire wingtip is yet to be fit-
... Such was the attention to detail that the
...onstructed P9374 even had to have its
...ery of eight Browning machine guns in-
...ed, although for legal reasons all of them
...iously had to be de-activated; Incredibly,
...guns are also fitted out with belted 1940-
...ed de-activated and thus inert ammunition;
...other unique feature to P9374 and the Spit-
... I were the retractable under-wing landing
...ps which are lowered by compressed air
... retracted by springs.

ing gear was sourced by the Duxford team and overhauled in house. Also, of course, there was a need to provide the correct tyres as we have already seen, and these were re-moulded, exactly as per original, by Dunlop the original maker.

In considering the final product, Martin Overall is proud of the masterpiece he has largely been responsible for creating, or at least assembling. With justifiable satisfaction he points out many of the bits recovered from the original P9374; a rudder pedal, the seat armour, cockpit cross-member heel bar and flare chute handle to name but few. He also highlights the modern spec electrical wiring, hidden inside its re-created 1940 period fabric braiding and, similarly, the linen-covered Bowden cable.

There are, though, a few modern and necessary compromises and one of these is the modern radio. This is neatly fitted inside the cockpit map box, which might almost have been made for this very purpose. With its lid closed the radio is completely hidden away, and its transmit button is also cunningly disguised on the port side of the cockpit. One other compromise is that the 1940 emergency CO_2 undercarriage blow-down bottle is replaced with a modern equivalent. Whilst the original bottle is fitted, it is just cosmetic although its handle is 'operational' and connected to a cable that operates a modern blow-down bottle fit-

Above: A new-fit modern radio was an essential compromise in the reconstruction of P9374, although it neatly slotted into the map box where it was hidden from view.
Below: The CO_2 'blow-down' bottle for lowering the undercarriage in the event of emergency and fitted on the starboard side of the cockpit.

The 12-volt voltage regulator was another masterpiece of engineering or, in this case, re-engineering. With only 24-volt examples available this one was re-wound and re-configured to working order as the correct 12-volt type.

ted well behind the pilot's seat.

Perhaps one of the very visible parts in the cockpit is the prominent 'biscuit tin' voltage regulator mounted just behind the pilot's head rest. Later Spitfires used a 24-volt system and a reasonable supply of 24-volt voltage regulators are still to be found, but Spitfire Is used a 12-volt system. Of 12-volt regulators, not a single one was to be found. Anywhere. As was often the case with P9374 it would have to be a matter of somehow creating one, or in this instance turning a 24-volt regulator into a 12-volt one. Externally identical, a 24-volt regulator was thus modified and re-wound by Guy Black's East Sussex aero engine workshops.

With the wings and fuselage completed, and the cockpit and instrument panel fit underway, it just remained to position the engine and propeller and to install the systems. First, we must look at the absolutely mammoth task of readying the engine and propeller assembly. And that was no small task.

10

THE ROLLS-ROYCE MERLIN III

The Rolls-Royce Merlin engine has sometimes been described as 'the engine that won the war'. Although perhaps an extravagant claim, it might well have some substance given its overwhelming importance to the Allied war effort and its widespread use in a variety of front-line RAF aircraft throughout the duration of the struggle. Notably amongst those, of course, the Spitfire and Hurricane were both Merlin-equipped and whilst the purpose here is not to give a detailed history of the Merlin per se, nor of its development, manufacture and usage, we need to look at the background of the Merlin's development, especially in the context of the Merlin III which powered the Spitfire I. It is also important to appreciate that the Merlin III is not an engine that is easily available in the twenty-

When the wreckage was eventually recovered during January 1981, the salvage team pulled the engine out of the airframe as they attempted to drag the Spitfire from the wet sand. The wreckage was taken to the nearby Calais Hoverport where this photograph was taken. It proved to be a source of a surprising number of components reusable in the engine rebuild for the project.

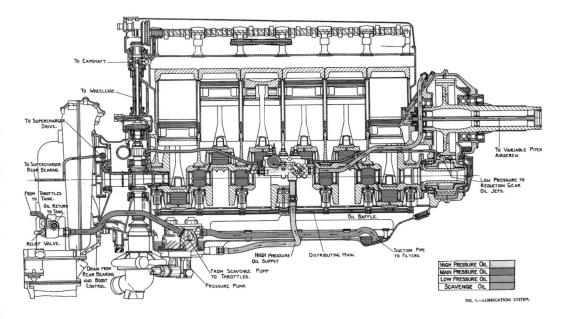

FIG. 1.—LUBRICATION SYSTEM.

HIGH PRESSURE OIL	
MAIN PRESSURE OIL	
LOW PRESSURE OIL	
SCAVENGE OIL	

first century; leastways, not in any running or airworthy condition.

Since one was originally installed in P9374 an example had to be found and restored, or built from near scratch, and to airworthy condition. The Merlin III was significantly different to other later Merlin engines, and it was not going to be a case of fitting any old Merlin into P9374, which would have been impossible within the demanding build spec, anyway. Retro Track & Air, based at Cam, near Dursley in Gloucestershire, a specialist aero-engine restoration company, rose to the challenge of constructing the engine for

Above and below: Oil diagrams from an original Rolls-Royce Merlin III handbook. When recovered, P9374's Merlin still contained the correct quantity of its original engine oil!

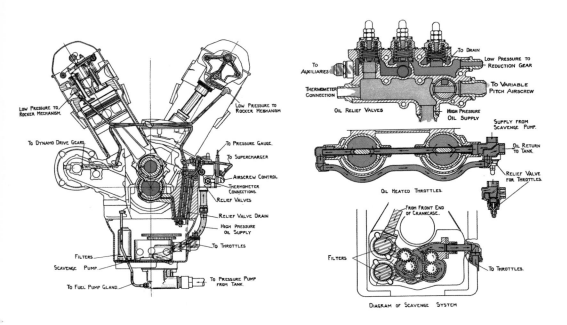

DIAGRAM OF SCAVENGE SYSTEM

P9374 and sourcing or re-manufacturing many components as well as incorporating a surprising amount of the *original* engine into the project. Given that P9374's Rolls-Royce Merlin had been immersed in sand and salt water for just over forty years, and subsequent to its recovery it had hardly been properly restored or conserved, this alone is a truly remarkable achievement.

The Rolls-Royce Merlin is a liquid-cooled V-12 piston aero engine, of twenty-seven litre (1,650 cu in) capacity. Rolls-Royce Limited at first designed and built an engine which was initially known as the PV-12, although this would later evolve and develop into the classic aero engine which today is as world-famous as the aircraft it would later equip – the Rolls-Royce Merlin.

In the early 1930s Rolls-Royce had started to plan its future aero-engine development programme and realised that there was a need for an engine larger than their twenty-one-litre (1,296 cu in) Kestrel which had been used with great success in a number of 1930s aircraft. Consequently, work was started on a new 1,100-horsepower aero engine designated the PV-12, with PV standing for private venture because the company received no government funding for work on the project at that time. The PV-12 was first run on 15 October 1933 and first flew in a Hawker Hart biplane (K3036) on 21 February 1935. The engine originally used the evaporative cooling system which was then in vogue but this proved unreliable and when supplies of ethylene glycol from the United States became available, it was adapted to use a conventional liquid-cooling system.

In 1935, the Air Ministry issued specification F10/35 for new fighter aircraft with a minimum airspeed of 310 miles per hour (500 km/h). Fortunately, two designs had been developed: the Supermarine Spitfire and the Hawker Hurricane although the latter was initiated in response to another specification, F36/34. Both were designed around the PV-12 instead of the Kestrel, and were the only contemporary British fighters to have been so developed. Production contracts for both aircraft were placed in 1936 and further development of the PV-12 was given top priority as well as essential government funding. Following the company convention of naming its piston aero engines after birds of prey, Rolls-Royce named the engine the Merlin.

Initially the new engine was plagued with problems, such as failure of the accessory gear trains and the coolant jackets and several different construction methods were tried before the basic design of the Merlin was set. Early production Merlins were also unreliable and common problems were cylinder-head cracking, coolant leaks, and excessive wear to the camshafts and crankshaft main bearings. However, it was in-service experience and operational demands and requirements that would lead to on-going developments of the engine to the extent that later variants were a far different, more reliable and much improved animal than were the early models such as the Merlin III as used in P9374. Nonetheless, the clock had to be turned back to 1939 and 1940 for this venture, and there was no question of using later engines to power this particular Spitfire. The Merlin III, and all of its known weaknesses, would be incorporated. First, let us look briefly at the developmental progress of the PV-12/Merlin engine.

The prototype and developmental engine types were the:

PV-12

The initial design using an evaporative cooling system. Two were built and passed bench testing in July 1934, generating 740 horsepower (552 kW) at 12,000 feet (3,700 m) equivalent.

MERLIN B

Two of these were built and an ethylene glycol liquid-cooling system introduced. 'Ramp' cylinder heads (inlet valves were at a 45-degree angle to the cylinder) were adopted. Engine passed type testing February 1935 and generated 950 horsepower (708 kW) at 11,000 feet (3,400 m) equivalent.

MERLIN C

Development of Merlin B but with crankcase and cylinder blocks now three separate castings with bolt-on cylinder heads. Engine first flown fitted to a Hawker Horsley on 21 December 1935, 950 horsepower (708 kW) at 11,000 feet (3,400 m) equivalent.

MERLIN E

Similar to C engine, but with minor design changes. Passed fifty-hour civil test in December 1935 and generated a constant 955 horsepower (712 kW) and a maximum rating of 1,045 horsepower (779 kW). The engine failed its military one hundred-hour test in March 1936 although it powered the Supermarine Spitfire prototype, K5054.

MERLIN F (MERLIN I)

Similar to the C and E engine and first flown in a Horsley aircraft on 16 July 1936. This became the first production engine and was designated the Merlin I. The Merlin continued with the 'ramp' head, but this was not a success and only 172 were made. The Fairey Battle was the first production aircraft to be powered by the Merlin I and first flew on 10 March 1936.

MERLIN G (MERLIN II)

This engine replaced 'ramp' cylinder heads with parallel-pattern heads (valves parallel to the cylinder) scaled up from the Kestrel engine. 400-hour flight endurance tests carried out at Royal Aircraft Establishment in July 1937. Acceptance test completed 22 September 1937. The engine was first widely delivered as the 1,030-horsepower (770 kW) Merlin II in 1938, and production was quickly stepped up as war clouds loomed.

The Merlin II and III series were the first main production versions of the engine type. The Merlin III was manufactured with a 'universal' propeller shaft, allowing either De Havilland or Rotol-manufactured propellers to be used and the first Merlin III was delivered to the RAF on 1 July 1938.

Perhaps no account of the Merlin engine would be complete without, at this stage, some mention of the fuel used to power the engines. From late 1939, 100 octane fuel became available from the United States, West Indies, Persia and also domestically. Merlin IIs and IIIs were adapted to run on this new fuel, using an increased boost pressure of +12 pounds per square inch (183 kPa; 1.85 atm). Small modifications were made to the engines which were now capable of generating 1,310 horsepower (977 kW) at 9,000 feet (2,700 m) while running at 3,000 revolutions per minute. This increased boost was available for a maximum of five minutes, and if the pilot resorted to emergency boost he had to report this on landing and have it noted in the engine log book. Using this boost was considered a 'definite overload condition on the engine' and the engineering officer was subsequently required to examine the engine and reset the throttle gate.

The foregoing is, then, a brief outline of the developmental path to the Merlin III as fitted in P9374. It was very much a recent engine design in 1939, and it was inevitable that in-service operational use would tease out problems and weaknesses that would ultimately be largely ironed out in its long production history. Like the Spitfire itself, the Rolls-Royce Merlin is considered a British icon and it was one of the most successful aircraft engines of the World War Two era, with many variants built by Rolls-Royce in Derby, Crewe and Glasgow as well as by Ford of Britain at their Trafford Park factory, near Manchester. The Packard V-1650 was a later and more developed version of the Merlin, and was built in the United States. Production of the Rolls-Royce Merlin engine eventually ceased in 1950 after a staggering total of almost 150,000 engines had been delivered, with the later variants being used for airliners and military transport aircraft.

In military use, however, the Merlin was superseded by its larger capacity stablemate, the Rolls-Royce Griffon, which went on to be fitted to later marks of the Spitfire. Incredibly, perhaps, Merlin engines remain in RAF service today with the Battle of Britain Memorial Flight and are thus the longest-serving engine in the air force, having been fully in continuous use for well over seventy years. That, though, does not make the acquisition of a Merlin III an easy task – especially not a fully serviceable one.

Peter Watts is proprietor of Retro Track & Air, a specialist large aero engine and historic racing car company based in Gloucestershire, which has a wealth of knowledge and experience of all things Merlin. Ask Peter most things you need to know about Merlin engines, and he will usually find the answer. It was only natural, then, that his company should be commissioned to rebuild a Merlin III engine for the P9374 project. And the brief was that it had to be the original engine from P9374. At least, that would be the starting point.

The full history of Rolls-Royce Merlin III (number 13769) is given at Appendix IV, and this engine had also been acquired by Thomas Kaplan along with the airframe remnants of P9374. Externally, and especially when it was first recovered, the engine had appeared remarkably sound and still retained its original black finish. This, however, was with the exception of the top end of the cylinder blocks, the rocker gear, cams and cam covers. All of these areas had suffered salt-water corrosion and the attention of souvenir hunters and vandals during the pre-recovery period. What had been beneath the sand, still cowled-up, had survived remarkably well, though. Post-recovery, however, the engine had not been either adequately preserved or properly conserved – leastways, not with any long term prospects of the engine being restored to run.

Internally, especially at recovery, it is very likely that many of the engine parts were in as-new condition. Almost certainly, the crankcase and sump still contained the original engine oil and the engine had been effectively sealed against any significant ingress of salt water or sand. Of course, the engine would have later dried out and had been subject to some interference and tinkering, probably to the detriment of internal components, in the years up to its acquisition by Thomas Kaplan. Externally, the impregnated salt water dried out leaving its saline crystal deposits in and on the engine casing. As the salt re-crystalised, so it began to take its toll. Gradually, corrosion began to set in causing significant pitting of the alloy castings and the rusting of steel components such that when the engine came into Peter Watts's hands a number of parts that would have otherwise

A rear view of the re-built Merlin III (number 13769) in the workshops of Retro Track & Air at Dursley, Gloucestershire.

been salvageable and re-usable were beyond saving. A particular misfortune was that the main crankcase, although visually and externally sound, had begun to degenerate badly in places due to the action of salt water such that it was no longer viable to use the crankcase in the project. That said, a surprising number of other valuable components were usable.

When 13769 went into the assembly shop at Retro, the stripped down and laid out parts yielded up the lower crankcase (more commonly although incorrectly known as the sump cover), reduction gearbox casing, the boost control and the relief valve assembly amongst other smaller but equally very important parts. From here, it was a case of Peter's team of skilled engineers constructing the engine from the ground up and incorporating all serviceable parts of the original engine into the rebuild. Although Merlin III engines are rare on a worldwide basis, John Romain of Historic Flying Ltd had sourced a potential serviceable core and this proved to be a valuable spares source. Added to that, Peter holds a significant stock of Merlin spares and many (although not by any means all) are interchangeable between various Merlin marks.

A close up of the Rolls-Royce embossed rocker box covers. The originals had been smashed by souvenir hunters in 1980.

In terms of a Rolls-Royce Merlin build, Peter Watts described the construction of P9374's engine to be something that was almost routine. The experience of his company made it so, although it is one of the earliest Merlin engines rebuilt anywhere in the world to airworthy standards. That, in itself, is a truly unique achievement and made it not quite so run-of-the-mill as the bigger Merlins fitted, say, to the Spitfire IXs that had almost become his stock in trade. With the Merlin III rebuild there were certainly problems, difficulties and new puzzles to be solved and overcome by the engineering team although they were but nothing when it came to comparing those issues to those that were yet to be addressed with the propeller unit to be hung on the front of number 13769.

Completed at Retro's Dursley workshops in Gloucestershire in the spring of 2010, the completed unit was transported to Cotswolds Airport (formerly RAF Kemble) in order to be mounted on the company's unique test rig. As Peter Watts explains:

"The test rig is a vitally important piece of kit in the ground testing of Rolls-Royce Merlin engines. It is a type of engine that overheats notoriously when ground-run in, for example, a Spitfire. This is because of the cooling problems inherent in the design associated with ground running and the reliance of airflow through radiators to keep the temperature down. Thus, ground testing a

Merlin fitted straight into a Spitfire is problematical to say the least and you only have a very short time before the engine overheats and boils. Then you have a problem. So, our test rig allows for a full package of testing regimes to be carried out as it is set up with an efficient cooling system to overcome this snag. Ours is the only test rig of its type in the UK, if not Europe, and is hugely beneficial – especially in test runs of engines like the Merlin III which was, to an extent, all a bit of a learning curve for us. We needed to be sure, before it was dropped into its engine mounts in P9374, that it would run to its operational limits and expectations and did everything it said it would do on the tin. It did!"

Thanks in no small measure to the skill of Peter's team it ran impeccably and was passed and signed off as fit to fly with a recommended life expectancy of two hundred hours.

Following on from Merlin 13769, Peter is now working on another Merlin III but this time for fitment into Spitfire N3200 for Mark One Partners and there is a likely expectation, too, that others will follow. Not least of all for one 'on-the-shelf' spare engine, but also one for the future project that is likely to see the re-creation of Spitfire I, P9373.

One last detail, though, had to be considered and that was the distinctive

Above: A rear-end view of Merlin III, number 13769. This photograph shows the supercharger and carburettor to advantage and illustrates the complex assembly of control linkages etc.
Below: A front-end view of the re-built Merlin engine for P9374.

ejector-type exhausts fitted to the Merlin III and the early mark Spitfires. Again, there is no supply of new-old stock and so they would need to be fabricated and for this Mark One Partners turned to Sven Kindblom in Sweden to manufacture the surprisingly complex designs from Inconnel, although when the bright and shiny silvery exhausts were first fitted there were concerns that they would not 'dull down' to the burnt bronze-brown hue that was their distinctive in-service feature. The concerns, though, were unfounded and after not many engine runs the tone of colour began to become more familiar and has continued to do so as the life of the engine progressed. As with all else, it had to look right and be right and the exhausts are, of course, a very distinctive and noticeable feature.

Left: The original brass plate from the supercharger unit of Merlin III number 13769.
Below: A final but essential detail was the construction of ejector-type exhaust ports for the Merlin. Without any plans or drawings to work from, these were skilfully crafted through a reverse-engineering process.

As an engine, however, it is true to say that the Rolls-Royce Merlin III is significantly, and quite naturally, far less reliable than the later marks of Merlin. For the owners, however, fitting one was essential.

The big day comes at last when the Merlin III is ground-run in Spitfire P9374 at Duxford.

THE DE HAVILLAND PROPELLER – MISSION IMPOSSIBLE

11

If re-building the engine and re-constructing the wings, fuselage and on-board systems had witnessed many obstacles, these paled when compared to the ultimate challenge; the propeller. The Spitfire I was fitted with an alloy-bladed De Havilland bracket-type propeller of which, quite simply, there were no available airworthy examples anywhere in the world. Some examples did exist, on static museum Spitfires, but even these were out of reach for the restoration team. Wreck-recovered examples were all beyond any practical use in terms of restoration to flyable condition, and only the fewest minor internal components might ever be usable after high speed impact with the ground. Although it had looked sound on recovery, the original propeller hub from P9374 would also have suffered considerable internal stresses and likely fractures and distortion through its impact with the sand – not to mention its long immersion in sand and sea water. Not only that, but the original assembly had disappeared in the intervening years. Also, and just like everything else with this project, the original type of propeller assembly was long out of production and so there was only one solution; make one. Doing this, though, was not exactly going to be a walk in the park but the owners once again turned to Retro Track & Air to solve the problem. And what a problem it was!

The first snag was that all engineering drawings and details pertaining to the De Havilland bracket-type propeller (specifically the DH 5/29 propeller assembly) had long since been disposed of when that company was subsumed by Hawker Siddeley Dynamics, and although the DH 5/29 assembly was a licence-built Hamilton propeller, that particular company evidently had no drawings either. Even if the De Havilland drawings did exist in archives with the Hamilton company, or with that company's successor, it is likely that product liability issues would prevent the release of such drawings to any third party. Moreover, the De Havilland licence-built version differed in a significant engineering sense from the US-built version as it adopted British specifications and, importantly, SBAC splines as opposed to the American AN spline. So, the lack of any drawings

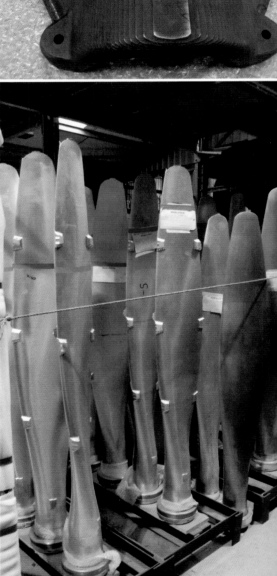

Clockwise from top left: The construction of the De Havilland propeller units from scratch by Retro Track & Air was a complex engineering task which began with these 'rough' forgings of the two halves of the propeller hub clam-shell design; Each half of the rough forging then had to be precision machined to the final finished version. Here, one of the hub halves is shown in a part-machined state; The rough and part finished propeller blades in the Retro Track & Air workshops. Like every part in this new-build assembly, each had to be forged and machined from scratch; Both finished hub halves are seen completed and assembled together. With no original or usable propeller assemblies of the correct type for P9374 available anywhere in the world, the only solution was to set up a plant to re-manufacture the units — the first time any had been made since 1940.

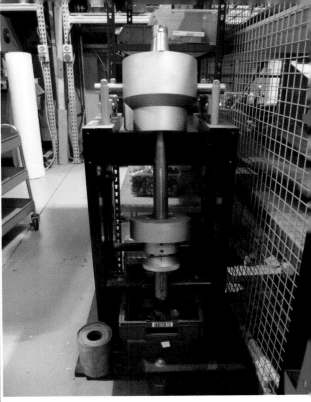

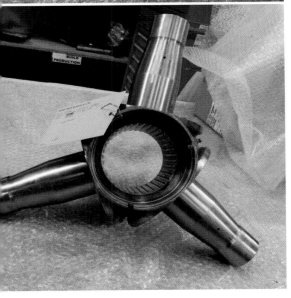

Clockwise from top left: The rough blades were ground down to their correct shape and profile, but using gauges at pre-determined station marks on the blades to ensure that the correct and accurate profile was achieved; Not only did Retro Track & Air have to build the complete propeller unit but they also had to design and build much of the engineering machinery for the process, including this propeller blade balancing machine; The internal precision machined 'spider' of the hub onto which each blade would eventually be slotted. The central aperture with its internal splines can be clearly seen; The propeller fitted to P9374 was known as a bracket type, as a balance weight moved up and down the bracket at the root of each blade as the pitch was changed.

was a major issue; no drawings usually meant no propeller. That, however, was not going to beat Peter and Stuart Watts and their team at Retro Track & Air.

Since no drawings could be located or accessed anywhere in the world it seemed that the only solution would be to reverse engineer the whole project. In other words, take a relic or museum condition propeller of the exact type and pattern, disassemble it, measure it, photograph it, draw it up and then design specific engineering plans to put this extinct propeller back into production. Furthermore, metal samples needed to be taken for laboratory tests to get the exact composition and type of metals used. To any layman, all of this sounds complicated in the extreme and any aero engineer will probably tell you that such a project would be almost impossible. The Retro team, however, were utterly determined to make the impossible possible. First, a suitable propeller needed to be sourced.

As we have seen from the opening paragraph of this chapter, getting one released from any museum collection was going to be extremely improbable. Inspecting, measuring and photographing such a propeller assembly might be one thing, but taking it apart for 'invasive' investigation was quite another. For one thing, getting the propeller back together *after* deconstruction might be problematical and much of the work required would be 'terminal' in relation to the continuing integrity of the whole unit. For instance, cutting cross sections (if required) would be impossible and taking metallurgical samples would be out of the question, too.

Clearly, another 'donor' propeller had to be found and Peter and Stuart Watts turned to the author to supply a relatively un-battered wreck-recovered example for use as their principal pattern. As it happened, the assembly was from another aircraft that had also been lost in northern France during the May/June 1940 period in what had been a high impact-speed crash, but it nevertheless made a perfect sacrificial unit that would at last enable this very ambitious project to get underway.

First, though, authority for the design and manufacture of the propeller assemblies had to be obtained from the Civil Aviation Authority. It was not as simple as just going ahead and designing and manufacturing a propeller for flight, and although Peter Watts and his team originally sought to manufacture these assemblies by special approval under their existing A8-20 certificate, the CAA ultimately decided that this complex engineering task for a high-energy assembly could only go ahead under a much more rigorous Design and Manufacturing Approval A8-21 certificate. Consequently, Retro Track & Air applied for and were granted the A8-21 certification and this, in itself, made Retro the only UK company in the historic aircraft industry able to carry out such design and manufacture. It also meant that the company were cleared for the design and manufacture of just about all and any aeronautical assemblies, structures, systems, engines and propellers for any non-certified and non-turbine aircraft. Thus, the official path was open for Retro to put back into production the 5/29 propeller unit.

Occasionally, a little bit of good luck assists any immensely tricky undertaking and it was to be the case here as, when investigating the metal type used in the propeller blade,

Peter Watts discovered that it was D.T.D 150 (superseded in 1966 by the designation D.T.D 150A), a light alloy specified for the manufacture of detachable blade airscrew forgings. By great good fortune it then emerged that D.T.D 150A was still in use and still available in 2010, and to the exact original specification.

Moreover an established aerospace forge-master was still able to produce the required blades – again, to the precise and original specifications. Mark One Partners had initially decided upon the purchase of seven complete propeller assemblies, a total of twenty-one individual blades being required and delivered as 'roughs' to Retro for finishing. As we will see, though, describing the process as 'straightforward' would be rather a slight on the engineering team.

The rough blades, straight from manufacture, first had to undergo a complex process of finishing. Initially, the propeller blades underwent rough machining but they had to be taken down to the final 10,000th/inch by orbital sanding machines progressing from rough down to fine abrasives. It was critical, of course, to obtain the correct and accurate profile to the blades, from tip to root, and this was done with the use of profile gauges placed around the blades at pre-determined 'stations'. The end result, a finished propeller blade, was then subjected to balancing; both static balancing and dynamic balancing.

In this process each blade is balanced to perfection by the insertion of lead 'socks' of varying weights that are buried inside the blade root and retained in place by tapered alloy plugs tightened into position by bolts. In order to carry out this crucial part of the process Peter and Stuart Watts and the Retro team had to design and build their own propeller-balancing machine, specific for use in the manufacture of what would be a limited number of Spitfire propeller blades. The task of designing and building the balancing machine was demanding and hugely expensive but it is illustrative of just some of the obstacles that had to be overcome in the overall process of reconstructing P9374.

Meanwhile, the complex hub assembly, now drawn and planned, was a further challenge. Described very simply, the hub unit itself is a clam-shell affair of two heavy steel halves bolted together by a total of six bolts, two in each 'shoulder' between the blade fixings. The blades themselves are simply clamped into place by a large flange at their foot, which is then retained inside a collar formed when each half of the clam-shell arrangement is bolted together. Inside this hub is contained the complex pitch-changing mechanism and the internal propeller 'spider'.

All of this had to be made from scratch. Unlike the situation with the blades, there was no engineering company in existence that was set up to make them, or who still had original tooling or dies. All of this, of course, had to be done from Retro Track & Air's plans drawn up from a relic 5/29 unit. Whilst RAF air publications and De Havilland manuals are plentiful and still exist to show how the units were put together (or taken apart, serviced and repaired) there was *nothing* available to show the detail of design, manufacture or construction of the one hundred and fifteen individual parts.

The two steel clam-shell halves of the hub, when originally manufactured, were made in a drop-forging process and, of course, the dies etc for this process, as well as the availability of the enormous drop-forge presses themselves, were not an option in 2010. To replicate this process the cost would have been quite astronomical to say the least, and although the overall cost of the project was huge anyway, a drop-forged propeller hub casing was simply not viable. Another way had to be found.

Consequently, and applying all of their vast engineering skills and experience, as well as the Civil Aviation Authority design and manufacturing dispensation, the Retro team designed and produced an alternative; the two hub halves would be forged from wrought steel and the roughs then machined down to create exact replicas of the originally manufactured hubs. Of course, these words cannot adequately convey the intricate process but suffice to say that the science, metallurgy and the design and engineering application that was necessary to produce the finished article was staggering. Remember, the whole propeller construction process had had to be reverse-engineered from the very outset and a thousand and one different problems solved along the way. Moreover, a production line had to be established on quite a small scale to produce units that were once produced in large numbers in vast engineering workshops that had been set up with huge presses and other plant designed to turn out units by the dozen, and on a daily basis!

With the two clam-shell halves machined and finished it was, of course, still necessary to design, manufacture and assemble all of the vital internals of the hubs; gears, thrust bearings, shims, shafts, splines and a plethora of other small components. All of this, including the manufacture of the balance weights and the balance bracket arms etc, was done in-house by the team. The bracket, of course, is an essential element of this propeller design, with the pitch being changed manually by the pilot operating a toggle in the cockpit. As he did so, the counter-weights mounted on the bracket moved a prescribed distance along the slots in the bracket assembly.

The biggest single internal component, though, was the 'spider' of the hub. Again, this was first produced as a casting which was then precision-machined down by Retro to the precise and accurate tolerances that had been measured and calculated from the relic donor hub. Again, to explain satisfactorily this process in any understandable lay-man's terms and within the remit of this book is nigh-on an impossible task, but suffice to say that the propeller is, overall, a supreme engineering masterpiece.

Assembled, it was of course necessary to ground-test the completed unit and this was first carried out fitted to a Rolls-Royce Merlin 500 mounted on Retro's test-bed truck and with the CAA present. It was then test-run on P9374's Merlin III, again at Retro's Kemble facility, before eventually being passed-off for delivery to Duxford in the early summer of 2010. The propeller, a re-creation of 1939 technology, had performed perfectly and within limits. The project was nearing completion.

Of course, not only was this old technology but it was also completely new territory for the engineers, for those who would go on to erect and maintain P9374 and, especially, for those who would ultimately fly it. After all, this was a very different propeller to those

Completed and newly delivered, the brand new De Havilland two-position propeller assembly has arrived at Duxford and awaits fitment to P9374.

which experienced Spitfire pilots in the twenty-first century were used to. It had very different handling and operating characteristics, too, and as we have examined some of the basic technical aspects and history of the Rolls-Royce Merlin III, so we should also study similar historical and technical aspects so far as they relate to the De Havilland 5/29 two-position manually-operated bracket-type propeller.

Originally, the Spitfire prototype (K5054) had been equipped and flown with a two-bladed fixed-pitch Watts wooden propeller (the manufacturing company name being co-incidental, and having no relationship to Peter Watts!) and this was still the standard fit when the Spitfire first went into production and squadron service with the Rolls-Royce Merlin II engine. Later, of course, the production Spitfire Is were fitted with the Merlin III and produced with the new De Havilland three-bladed propeller (as per P9374) and this was a vast improvement over the fixed-pitch two-bladers. The De Havilland unit had two settings, fine for take-off and coarse for top performance. It worked well until a pilot forgot to select fine pitch for take-off, and this was a cardinal error for many nascent Spitfire pilots of the period – notably Douglas Bader who had just such an incident at RAF Duxford, P9374's twenty-first century home. In his biography *Reach For The Sky* the story is told:

> "Quickly he strapped his straps and pressed the starter button; the still hot engine fired instantly and he was still winding his trimming wheel as the plane went booming across the grass. The other two Spitfires were shooting past him,

pulling away, and he sensed vaguely at first, and then with sudden certainty, that his aircraft was lagging. A quick glance at the boost gauge; the needle was quivering on 6 ½ lb – maximum power. She must be all right; but she was still bumping over the grass, curiously sluggish, running at a low stone wall on the far side of the field. The fence was rushing nearer, but she still stuck to the ground. He hauled desperately on the stick and the nose pulled up as she lurched off at an unnatural angle, not climbing. His right hand snapped down to the undercart lever but almost in the same moment the wheels hit the stone wall and ripped away. At nearly 80 mph the little fighter slewed and dipped a wing-tip into a ploughed field beyond; the nose smacked down, the tail kicked up – she nearly cartwheeled – the tail slapped down again and she slithered and bumped on her belly with a rending noise across the soft earth.

"The brain started working again and began wondering what had happened as he sat there with everything so suddenly quiet he could hear the silence and the hot metal of the engine tinking as it cooled. Automatically his hand went out and cut the switches and then he was motionless again apart from the eyes wandering round the cockpit looking for the answer. It stared back at him – the black knob of the propeller lever on the throttle quadrant poking accusingly at him, still in the coarse position.

"His stomach turned. Oh, hell no, not that classic boob! He couldn't have. But he had. Angrily he banged the knob in."

Unique to the Spitfire I was the propeller pitch knob and this is the original control knob from the wreck of P9374.

Quite apart from the cockpit workload issues, and setting aside its improvement over the fixed-pitch two-bladed propellers, it was soon clear that the De Havilland two-pitch manually operated unit needed yet further improvement. Consequently, on 5 April the Air Ministry asked De Havillands if it would be possible to convert the propellers to constant-speed units of the type that were already fitted to many multi-engine aircraft then in service. De Havillands were not overly impressed with the suggestion, and although they offered alternative solutions nothing further happened until 9 June 1940 when an RAF engineer officer contacted them again to ask if a propeller on one of his Spitfires could be converted to constant speeding "without a lot of paperwork and fuss".

It could, they said, and the process of converting all of the existing units was set in train, as well as the implementation of the constant-speed unit in future production Spitfire Is.

However, and because of the somewhat 'irregular' manner in which this process had been put in train there were apparently protracted problems between the Ministry of Aircraft Supply and De Havillands. It is said that the company were never properly paid for the work, and legend has it that a clerk who worked for the De Havilland company commented: "We shall probably never get paid for this work," to which his colleague is said to have responded, "Well, if it doesn't get done we may never live to be paid for anything."

In-service Spitfires were all to have been converted to constant speed by 20 July by teams of De Havilland engineers working with RAF squadron fitters and it was reported on 16 August 1940 that all Spitfires in squadron service or held in storage had been converted to constant-speed units in accord with Spitfire I modification number 76 of 3 July 1940. To an extent, the foregoing has little bearing on P9374 as she was in May 1940, or how she has been re-constructed, but it is significant in understanding the intricacies of the De Havilland two-position propeller unit as built for and installed in the aircraft, just as it had originally been. Worth mentioning, too, that other early Spitfire I aircraft that are currently under re-build or re-construction for other owners will neither have the 'original' fit of a Merlin III or the correct De

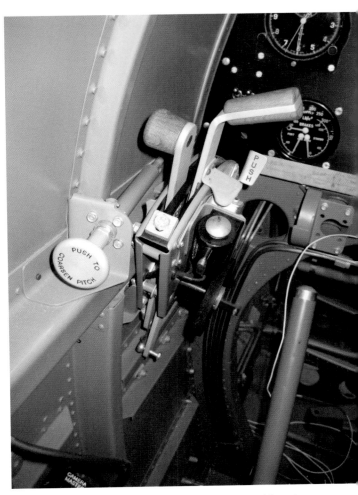

The pitch control knob is shown in place and fitted into the cockpit of P9374 to the left of the throttle quadrant.

Havilland propeller units but have, instead, utilised substitute or compromise solutions with both non-standard Merlin engines and propellers. For P9374 no such compromises were ever contemplated.

Whilst the construction of the new-build propeller units, headed by Stuart Watts, was a very rewarding project, Peter was clear that the phasing out of the old units in 1940 and their replacement with the constant-speed units was both timely and necessary at that crucial period. "Had the conversions not taken place," remarked Peter, "then the outcome of the Battle of Britain might well have been different given the improvements to RAF fighters' operational performance that the constant speed units had provided." His views echoed the sentiments expressed in the conversation between two De Havilland clerks seventy years previously. Peter, and the De Havilland clerk long before him, are probably not wrong in their assertions. For example, it is recorded that the constant-speed conversions resulted in a much improved rate of climb in the Spitfire I, and gave a surprising extra 7,000 feet of altitude. Additionally, it was noted that the new constant-speed units gave an improved manoeuvrability at altitude, a reduced take-off run and, overall, they were reckoned to have an efficiency of 91%.

No such niceties as constant-speed units, though, for P9374 which must quite rightly soldier on with her 'pre-modification 76' propeller. It is, though, an assembly that is very much the crowning glory of the entire project. The metaphoric icing on a very lovely cake.

Whilst the engineering and construction challenges of this project resulted in the re-creation of Spitfire I, serial number P9374, the culmination of that work achieved the end result of a perfect Spitfire of the period and faithful in every respect to a production model produced during 1939/1940. Except, of course, for one thing when it finally left the engineering jigs; the paintwork. In fact, and in many respects, the finishing of this aircraft presented its own unique challenges and resulted in yet another investigative detective trail led principally by Historic Flying's Col Pope, an old hand at researching the finish and paint schemes of wartime aircraft and a specialist in his art. Again, one of the difficulties facing Col was that there was a limited amount of data to go on and aside from photographs and drawings showing the period paint scheme it was sometimes surprising how little information existed, for example, regarding the wording of the numerous stencilled instructions across the airframe.

The small number of Spitfire I aircraft still extant no longer had wholly original paintwork and thus were of limited research value and therefore much of it had to be worked out. By great good fortune, Peter Arnold's discovery of the photograph showing the beached Spitfire provided a sound basis for the camouflage pattern and, importantly, the

fuselage markings including the style of roundel, fin flash and identification letter. Without that knowledge, important aspects would have had to be guessed at, although with care and attention and Peter's invaluable photograph Col was able to work out precisely how P9374 was finished. Ultimately, its unique and unusual paint scheme was replicated

Although it is hard not to conclude that this must be a twenty-first century re-enactor alongside a restored Spitfire it is, in fact, an original 1940 period photograph of a pilot standing by his aircraft. The distinctive period camouflage finish seen here had to be replicated for the P9374 reconstruction.

113

As the paint finishing progresses, so the serial (P9374) is marked out for painting although the yellow outer ring has not yet been applied to the fuselage roundel.

on the airframe in such a way that it undeniably returned to P9374 its true identity.

When it came to the point in the project for the application of an authentic colour scheme, a considerable amount of extremely detailed research was required and the brief passed to the team from the owners was quite simple: "Put it back to exactly how it was when it crash landed on the beach." This gave both a fixed date and starting point, and the availability of black and white images of the Spitfire at the beach location further assisted the process. Chris Norfolk, of Historic Flying, was appointed in overall charge of the colour scheme and its application, whilst Col Pope was asked by John Romain to act as an advisor to the task because of his in-depth research of previous schemes and his knowledge of the scheme fluctuations that occurred during the period.

The images available indicated that the standard Ministry of Aircraft Production Pattern No.1 for Single Engine Monoplanes had been applied, and that the 'A' scheme was used. 'A' and 'B' schemes related to the actual layout of the camouflage pattern on the airframe. On the wing surfaces they were a basic mirror image of each other and on the fuselage they were easily distinguished by the direction of the camouflage pattern striping. For example, on an 'A' scheme the port side stripes of dark earth and dark green would slant to the back of the aircraft whereas on the 'B' scheme they would slant towards the front. At this point in the production of Spitfires there was a policy of most aircraft with serial numbers ending in an even number having the 'A' scheme applied, and those with an odd serial having the 'B' pattern. However, this was not set in stone and some production batches had this general policy reversed. These batches are known and recorded, but

The completed fuselage has now been fully painted in authentic period markings at Duxford, an airfield which saw the first Spitfire Is in service during 1939. The wooden stepped staging is to allow engineers easy access to the fuselage which is still being fitted out with the various systems under installation.

with the photographic evidence available it could be determined that P9374 was certainly in the 'A' scheme. Colours applied to RAF aircraft were determined by the policy in force at the time and were not random, notwithstanding that the scheme might seem so to the casual and uninformed observer. There were strict conventions, and all of them laid down by specific Air Ministry orders (AMOs).

With P9374 it was clearly without doubt that the top surface camouflage was of the dark green/dark earth specification as drawn up in Air Diagram 1160 of 1937, however the lower surfaces of this particular Spitfire were a different story! At this point in the research things all became very complex. There were considerable variations of lower surface colour schemes applied during 1940 and as P9374 crashed on 24 May it preceded the Air Ministry Signal X915 of 6 June 1940 that instructed all units to apply 'Sky type S' to the undersides of all aircraft and remove the roundels. This eliminated the possibility that the undersides were painted 'Sky' and the inherent variations of shade in that particular colour. Attention was now turned to the distinctive night/white, night/silver or night/white/silver patterns often seen in aircraft of the period. The pictures on the beach showed the aircraft partially submerged and were of no possible help to the team. The basic night/white combination was introduced following major problems with identifying friendly fighters from the ground and the likelihood of anti-aircraft batteries opening fire on friendly aircraft. In addition, the early Chain Home radar coverage was non-effective inland and so the Observer Corps

This view shows to good advantage the distinctive half white/half night black under-surface paint finish.

was essential to identify visually aircraft and their friendly or hostile nature. This distinctive black/white/silver underside scheme (or variations thereof) was undoubtedly an aid for the Observer Corps in at least determining nationality if not the specific aircraft type.

In May 1937 RAF Fighter Command had made a recommendation to the Air Ministry that the underside of one lower main-plane should be painted in silver dope, and the other in dull black so as to provide a distinctive visual identification facility. Trials were carried out at North Weald with biplane fighters having one wing black and the other white. Although there were mixed results against cloud or direct sun, overall it was found to be highly distinctive against clear sky or scattered cloud. After small-scale introduction of the pattern on some units, it appears that the September 1938 'Munich crisis' prompted the Air Ministry to authorise Fighter Command's requests to roll the scheme out across its entire fleet. Spitfires thus began leaving the Supermarine works in the new black and white scheme in April 1939.

It should be pointed out that the colour used was actually described as night, and never black. Though perhaps sounding like the same shade, they are in fact noticeably different. Night has a charcoal grey depth to it, as opposed to the intensity of true black. Thus, and having now narrowed down the colour scheme layout, the team had to decipher whether the lower surfaces were simply night/white or the often seen night/white/silver layout. In the latter pattern, the lower engine cowling and rear fuselage aft of the wing are seen to be in a basic aluminium silver dope finish, with only the wings and tail surfaces finished in night/white. With no images of P9374's lower surfaces available it fell to studying contemporary aircraft serving on 92 Squadron at this precise time so that any available evidence could be gleaned. The aluminium finish was certainly seen to be on some machines, but in large part the two-pattern night/white layout was predominant, and with this evidence

it was decided to adopt this latter scheme for P9374. Having sorted out a decision on that part the next question arose; were there any roundels on the lower wings?

On 15 May 1940 RAF fighters were considered highly likely to be operating over France in support of the British forces then being pushed back to the coast. Signal X296 was issued that day ordering the application of lower wing roundels to all home-based fighter aircraft. They were to consist of a red centre, surrounded by rings of white and of blue, both of equal width, and to be as large as possible without straying onto adjacent aileron surfaces. This was further modified on 4 June 1940, with signal X479 instructing that the underside roundel on the wing painted in night was to be encircled with a yellow band 'of convenient width', but not less than one quarter and no greater than the full width of the blue band. This was to be carried out at the earliest possible moment.

The under-wing roundel on the night black (port) wing. In this image some of the many stencilled servicing instructions can be seen, in this case showing where the under-wing trestle support should be positioned. Each set of instructions had to be carefully researched and correctly applied.

However, and as with other modifications, there is evidence that this may have been already put into effect prior to the date the signal was actually issued and that operational squadrons were clearly applying the change 'in the field'. Therefore, the signal simply made it official across the entirety of RAF Fighter Command. It was thus decided that the yellow band should be applied to the resurrected P9374, although naturally it is recognised that this may be in error. It is, though, a best guess – albeit an educated one.

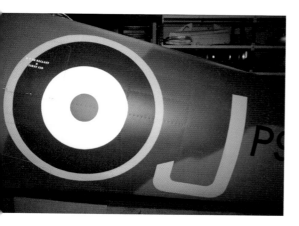

The fuselage roundel with its unusually proportioned outer ring of yellow and the individual code letter J. It was possible to scale and measure these markings accurately from the photograph taken of P9374 on the beach in 1940.

Perhaps the most visibly significant marking on this aircraft was the fuselage roundel. P9374 would have worn the standard red/white/blue layout during its short time on 92 Squadron, although photographic evidence exists to show that at least P9371 initially had a blue/red roundel and without any white rings. Evidence on wreckage discovered in the remains of P9373 also indicated that this Spitfire had once had the same blue/red fuselage roundel, too, and so it is not unreasonable to assume that, maybe on delivery, P9374 also originally had this same style of fuselage roundel applied. Trials to improve further the recognition of friendly aircraft had, though, been continuing and had already seen RAF Coastal Command carrying out experiments during February 1940 when an Avro Anson was painted with a yellow band surrounding the fuselage roundel. The results were very favourable, and on 1 May 1940 the Air Ministry sent signal X485 to all commands instructing that yellow bands the same width as the existing blue band were to be applied outside of the blue outer ring of the roundel. In addition, vertical stripes of red, white and blue, each of them to be the same width, were to be placed on the fins of all RAF aircraft. No specific sizes were issued at this stage, and this consequently caused chaos amongst various squadrons as they struggled to implement the new requirements and to paint up their aircraft appropriately during a time of war.

The details for the application of the fin stripes only stated that the blue section should be adjacent to the rudder on both sides, with this later being amended to include the proviso that the stripes did not have to cover the entire fin area. In respect of the fuselage roundels, many fighter aircraft found themselves with huge roundels almost literally falling off the fuselage sides after the yellow was applied since they were simply not physically big enough to accommodate the new expansion of roundel size. Signal X740 was then issued on 11 May to try to clarify the confused situation.

It stated that slim fuselage aircraft were required to have the whole roundel reduced in size, so as to accommodate the increased area requirement for the yellow band. However, due to operational demands, and merely as a temporary measure, the yellow band could be applied in a reduced thickness. This led directly to a huge variety of sizes and layouts, even within a single squadron, when aircraft were delivered as attrition replacements from MUs or other non-combatant units. Since the photographs of P9374 on the beach showed the roundel in clear detail, the team were able to reproduce exactly what it had worn on the fuselage side on 24 May 1940. Thus Col Pope used a method of scaling from the photographs to reproduce the exact proportions of the fuselage roundel, after which the yellow ring was found to have been two inches wide. This made the overall size of the roundel thirty-nine inches. So, whilst the roundels might look odd and of incorrect proportions to those with at least a passing knowledge of 1940 RAF colour schemes (and particularly to aircraft modellers!) they are absolutely correct to the colours actually worn by P9374 on that day in 1940.

The fin flash was a simpler issue, as the scaling clearly showed it was of what might be described as standard proportions and conforming to the Air Ministry Order X 485. Now, the main scheme was sorted out but some of the all-important detailing had to be tackled.

The Squadron code letters (or lack of) were to cause a huge debate amongst both the team and some of the project's outside advisors and consultants. 92 Squadron wore the identification codes GR at this time but at some point between June and August these were changed to QJ although no certain or absolute date for this has been established. *(Note: to confuse matters further, and much later on in 1940/1941, it emerged that 92 and 616 Squadrons, both of them operating Spitfires, were using the QJ codes.)* There was much keenness by some of those involved in the project to apply the full codes GR-J, as it had been established beyond doubt that the aeroplane had worn them at some point. A school of thought still exists with some that GR-J should be re-applied, although this would remove from the scheme the originality for 24 May 1940 which had been the initial brief and intention. Thus, the painting team stuck to their guns and followed the owner's brief to recreate how the Spitfire was when it had settled onto the sand.

The picture taken of P9374 on the beach clearly shows no code letters forward of the roundel, and a closer examination of the photograph reveals what *might* be traces of scuffing or of an abrasive removal (or even a rough over-painting application) in the area where the GR codes would have been applied between the cockpit rail and forward of the roundel. The team, and Col Pope, took the collective view that the aircraft was perhaps in the midst of having the codes changed from GR to QJ when it was required for operations, and it therefore took off on 24 May marked just as J.

The view of this author, however, remains that the code letters were indeed removed, but this was when the Spitfire had been allocated for special escort duties to cover the Flamingo aircraft that had been taking Churchill to Paris for talks. This escort flight never occurred (as we have already seen) as P9374 was later ferried back to its Northolt base before resuming operational service shortly thereafter – and presumably before any squadron code letters could be re-applied to the aircraft. In the fullness of time further research may yet reveal the truth, but both sides of the argument concluded that without any doubt the GR codes were not on the aircraft when it had crashed. And that was all that mattered. But what of the individual code letter, J?

Each squadron aircraft within the RAF was allocated an individual code identity letter, and on 92 Squadron P9374's was J. The style of the letter itself was not visible on the photographs in respect of its lower portion because the aircraft had sunk into the sand, but it could be seen to equate to the layout of how and where a J would sit on the fuselage. The single vertical stroke told the team that this must be a J, and this was corroborated by a photograph of an aircraft in 92 Squadron service at this time and coded GR-J. It can only have been P9374. Col Pope also found out that other 92 Squadron aircraft seemed to have an unusual hooked barb on the end of some of the letters they used, as well as having an unusual style to the letter G in the squadron code. He recognised the hooked barb as that of a font used pre-war by the RAF, particularly on biplanes, and he was able to supply the correct style of J from his own research and information archive. It may be assumed that the sign-writer on 92 Squadron (probably a rigger with the best painting aptitude!) was most likely a regular who had long service and, as such, was using a style that he was both familiar and comfortable with. Code letters were not regulated in respect of style at this point in time, and throughout the war a huge variety of fonts were to be seen even after strict guidelines had been laid down by the Air Ministry. Scaling of the photographs revealed the letter dimensions were a stroke thickness of four inches, with a height of thirty inches and width (across the body of the letter) of twenty inches. The letter was painted in the specified code letter colour in use at this time; medium sea grey. The overall scheme was now finalised.

Stencil details were decided by Chris Norfolk of Historic Flying Ltd (HFL) but there was no absolutely set instructional or servicing stencil plan, and machines differed considerably. Some instructional markings were mandatory and were laid down by the Air Ministry and the aircraft manufacturers had resident Aircraft Inspectorate Division (AID) personnel checking that these were applied rigidly and as per drawing. For instance, a 'Trestle Here' marking sloppily applied in the wrong place under a wing could have severe and unfortunate consequences if the instruction was followed through and complied with during squadron maintenance.

HFL therefore applied a variety of stencils to P9374 representing the most likely to have been worn in service at this point. Some manufactured aircraft are known to have had serial numbers of individual major components stencilled onto that component, or else sub-contractor inspection stamps and identification marks, whereas other identical machines only had instructional details. Paint specifications were applied as a stencil on wartime aircraft to indicate the paint type used. These were replicated on P9374, with the fabric-covered surfaces being DTD 83A and the metal surfaces denoted as DTD 314. Modern paint was actually used, though, which is to a very much higher quality than the wartime material specifications. HFL usually paint their projects as unassembled aircraft (ie with wings and fuselage painted separately before erecting as a complete aircraft), and this was the case with P9374.

The fuselage was painted first, in December 2008, with the application being jointly carried out by Chris Norfolk and John Loweth of HFL. When the wings were painted much later on it was decided to follow a precedent set by Col Pope on his scheme for the Battle of Britain Memorial Flight Spitfire IIa, P7350, which was painted to represent a Spitfire I, serial number R6760. On this scheme, a gas detection patch had been in-

The port wing viewed from the cockpit showing the yellow gas detector diamond-shaped patch and various wing walkway markings.

corporated on the port wing top surface. These were common on some squadrons between January and August 1940 and consisted of a patch of a detector material, or possibly sometimes of gas sensitive paint, that was applied to the top surface of the port wing and set six feet and six inches from the centre line of the fuselage.

The idea was simply that if enemy forces used a chemical agent, such as mustard gas, then the upward-facing detector patch would register the deposit of this agent causing a chemical reaction to occur and thereby change the patch from a greenish yellow to black. This would indicate a gas attack to ground crew and they could accordingly 'mask up' with their respirators to avoid incapacitation. How effective this would have been is questionable because the ground crew were more likely to already be suffering the effects of the gas by the time they had noticed any colour change. All the same, P9374 would probably have had the gas patch in May 1940 and therefore it was decided to apply it.

Despite exhaustive research it could not be determined what the patch was really made of, or applied with, but Col Pope concluded that it was most likely a material such as an impregnated paper which had then been pasted onto the wing and with a sealing bead of red dope applied at the edges to prevent airflow peeling the patch from the wing. Most patches were upended squares, applied to present a diamond shape, and the common size was eighteen to twenty-four inches. The red dope edge was noticeable in photographic archive material. A suitable colour had been determined during the painting of P7350 in the R6760 colour scheme, and this was duly replicated on P9374 as a painted square, since the application of a stuck-on patch would lend itself to wear, tear and slipstream damage. To represent the doped sealing edge, a one-inch band of wartime roundel red was overlapped by half an inch onto the yellow area. As a point of interest, the same patches were also noted during the period to have been applied to airfield ground equipment like fuel bowsers and trolley accumulator starters as well as to aircraft. Obviously, this was at a time when it was felt that the enemy might well use poison gas and concerns about this eventuality were running high.

The only other major marking application to both wings was, of course, the standard upper-wing RAF roundel of the period in a standard blue/red. As a regular pattern design it was, perhaps, the least contentious and most easily clarified marking on P9374, even though no sign of it can be seen in the beached wartime view of P9374. Whilst on the subject of the wings, however, it should be remembered that as a Spitfire I, P9374 was fitted with fabric-covered ailerons. Some early examples of the Spitfire I are known to have un-camouflaged ailerons, with these control surfaces being left silver-doped. The rationale for this is believed to have been a concern that painting the control surfaces might well throw them out of balance or trim, but there is no reason to suppose that P9374 had anything other than fully camouflaged ailerons on the upper surfaces, and with their undersides either white or night as appropriate to the wing.

Above: A detailed refinement was the addition of the white painted serial P9374 on the tip of one propeller blade.
Below: This detail from the 1940 image shows evidence of the aircraft serial number painted on the propeller blade tip. This was relatively unusual, but has been seen as a feature on a number of period photographs of Spitfire Is.

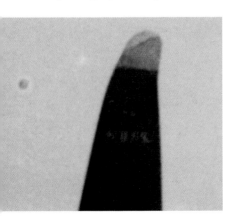

Other individual details appertaining to the finishing of P9374 included the hand application of the aircraft's individual serial number to one blade of the propeller as this had been noted on a photograph of another Spitfire I and was rather liked as a detail by the HFL team. Since it also *appeared* to be present on P9374 there was a *prima-facie* case for adding it. However, some debate continues with the P9374 team; does the photograph of P9374 on the beach show a painted serial on the blade just below its yellow tip or is this merely a scuff mark caused when the end of the propeller blade had 'kissed' the sand? Another important detail is that the undercarriage assemblies are 'handed' to the related white/night underwing paint scheme and painted accordingly; landing gear legs, wheels and door interiors.

Using all of the available evidence, both photographic and archival, the HFL team have done a thorough job in recreating the Spitfire exactly as it was when it crashed on the beach at Calais on 24 May 1940, not only in its construction and fit but, most importantly, in its finished paint scheme. Some aspects of the paintwork worn that day could not be fully authenticated and confirmed 100%, but that is not to say there may yet be more evidence out there. As such, the finished job might certainly be open to debate by serious researchers and specialists, but it is to be hoped that any such discussion simply throws up good historical evidence and clues to assist future projects or else in correcting any missed or unknown detail on P9374 rather than by fuelling unwarranted criticism as regards the paint scheme accuracy.

There can be no doubting that the Historic Flying team and Col Pope have pulled off a remarkable achievement in clothing P9374 just as she was. Look at the Spitfire today, and more than any other preserved or restored example P9374 transports the beholder right back to 1940.

Ready to go! Looking resplendent in its new markings and with every last detail now completed, the modern P9374 waits in its Duxford hangar for its first flight – seventy-one years since the original came to grief in France.

13 INTO THE AIR

With the metamorphosis from wreck to beautiful aircraft finally completed during the summer of 2011, the worldwide community of historic aviation enthusiasts waited for the final emergence and flight of Spitfire P9374. Spotters and photographers lurked impatiently at the eastern end of Duxford airfield for some good few weeks when it became apparent that the Spitfire was all but ready and its first test flight must be imminent. Occasionally, P9374 was wheeled out of the HFL hangar for engine runs and other work

Below: The triumphant team of dedicated engineers, technicians and support staff of Historic Flying Ltd at Duxford who have overseen the project to bring P9374 back to life, headed by John Romain (front centre) who also test flew the aircraft. Included in this shot is owner Thomas Kaplan (front left) and his colleague Simon Marsh (front right).
Right: Spitfire I, P9374, stands ready for its first flight on 1 September 2011.

John Romain, Director of Historic Flying Ltd and test pilot of P9374.

preparatory to its first flight, thereby causing flurries of excitement amongst the waiting enthusiasts but only for that excitement to be dampened when the engine was finally shut down and the Spitfire wheeled back inside again. Finally, on 1 September 2011, everything was in place and ready. The late afternoon and early evening weather was perfect and all was set, at last, for the momentous first flight. HFL's John Romain, who had overseen the reconstruction at Duxford, was naturally the test pilot for this epic flight. Here, John takes up his own story of that milestone event:

"Looking through my logbook reminded me that I have test flown quite a few Spitfires. The types vary and include both Rolls-Royce Merlin and Griffon-engined examples. Some were 'clipped wing' models, and two are Mark IX two seaters.

"In all cases, though, the various Spitfire marks had similarities. They all had an engine-driven hydraulic system to retract and lower the undercarriage. All were fitted with constant-speed units to control the propeller RPM and all had the later metal-covered ailerons.

"P9374 has been restored to exacting standards and is a true Spitfire I. For me as the test pilot this gave a few areas of discovery over the 'standard' Spitfire air test. No Spitfire I fitted with all the early systems, including engine and propeller, has flown since the end of 1940. Any aircraft flying after this time had been 'field modified' to enhance their modification state.

"The aircraft is powered by a Rolls-Royce Merlin III and has a two-pitch De Havilland propeller. The early engines were not the 'power horses' of the later marks and this example gives 940 horse power at +6 ¼ pounds of boost. They are also delicate engines to handle as the early one-piece cylinder heads and banks were particularly prone to cracking and losing coolant.

"Having a two-pitch propeller creates its own problems, too. There is no constant-speed unit to govern the engine RPM and therefore the pilot becomes the governor, constantly monitoring RPM against boost or throttle settings. Any speed change requires, in addition, an RPM change.

"The airframe is simply the earliest Spitfire design. I expected the controls to be quite light and responsive, thinking that a comparatively light aircraft and early design would probably reflect Mitchell's intent closer than the later types.

"Undercarriage retraction is a manual, hand-pumped affair. Testing in the hangar showed this to be a very simple and easy operation. Whether this would change with the aircraft airborne and having air loads applied to the landing gear I did not know.

John Romain, a very experienced warbird and Spitfire pilot, settles into the cockpit of P9374.

"The ailerons are fabric-covered rather than the later metal-covered assemblies. This in itself I did not expect to give problems, but I had asked veteran Spitfire test pilot Alex Henshaw about them – just in case. He simply asked how fast I would intend flying! 'Anything up to 400 MPH is fine John, over that they do get a bit heavy'.

"Reading the quite sparse Spitfire I pilots notes did not show any areas of additional concern, and so I prepared for the first flight with a balance of excitement and trepidation.

"Walking up to P9374 really impresses on you the fact that she is a Spitfire I. The straight legs of the undercarriage, the balloon tyres and 'baby' engine/propeller combination are striking features of the many wartime pictures I had seen of Spitfire Is on the grass at Duxford.

"With full fuel, my own weight and the parachute, the aircraft weighed in at 5,991 lbs and the centre of gravity position giving 6.6 inches aft of datum. She was not too far from the wartime operational weight of 6,200 lbs. Strapping in was carried out easily and it was not long before I was at last on my own and in the 'office'. Taking a few minutes to settle down I carried out my normal left to right check of everything.

"Cockpit side door closed and locked.
Rudder trim tab at neutral.
Elevator trim tab, one division 'nose up'.
Propeller pitch control at coarse.
Throttle, full and free and closed.
Mixture control – auto rich.
Throttle friction adjusted.
Electrical power switch – 'on'.
Undercarriage indicator light switch – 'on'.

Contact! In a cloud of blue exhaust smoke the Merlin engine roars into life before its first post-reconstruction test flight.

Magnetos – 'off', starter magneto – 'on'.
Radiator door – open.
Flap control – up, and air pressure – checked.
Flight instruments – checked.
Engine instruments – checked.
Fuel selector – 'on'.
Undercarriage selector at 'down' position.
Seat height – adjusted.
Headset lead – connected.
Canopy – free to slide and open.
Flight controls – full and free.
Brakes – on and checked – air pressure – check.

"With that done I am ready for engine start. The engine had been run earlier and so was still quite warm. If it had been cold then a good five loaded primes of fuel would be required. However, when hot or warm much less priming fuel is needed. I therefore gave three strokes of the Ki-Gas primer pump and then screwed it back into the closed position. Opening the throttle just half an inch I checked brakes 'on' and with the stick hard back call out: 'Clear to start!' With

John Romain cautiously taxies P9374 out before take-off on its first flight. Time spent on the ground with the engine running is extremely limited if overheating is to be prevented.

the propeller turning through its third blade the engine surges into life with a cloud of oil smoke and exhaust. What a lovely smell! Somehow, you never get tired of starting a Merlin engine.

"The oil pressure is instantly checked and is sitting happily at 60 psi. Selecting the propeller from coarse to fine position sees the oil pressure drop slightly and then return to normal. The engine is now running very smoothly and all the engine instruments are showing normal readings.

"Due to the engine already being warm I am conscious that I do not have a lot of time before take-off. The coolant temperature will rise quite quickly and so with that in mind I wave away the chocks and, releasing the brakes, start the taxi out to the runway.

Directional control is good and only small amounts of brake are necessary to weave from side to side on the way to the runway. Temperatures are rising now towards 95°C on the coolant – so, not too much time left! Stopping near the end of the runway I quickly go through the pre-take-off checks:

"Trims – set.
Throttle friction – set.
Propeller – fine pitch.

The moment of take-off as a newly re-born P9374 gets air under its wheels for the first time during the early evening of 1 September 2011. It was a moving event and a crowning achievement for all of those involved with the ambitious project.

Magnetos – 'on' starter magneto – 'off'.
Flaps – 'up'.
Radiator door – open.
Fuel – on pressure 2 ½ psi.
Instruments – set and checked.
Controls – full & free.
Harness – tight.
Canopy – open and latched.

"I leave the engine run-up until last as this will warm up the coolant even more. So, with the stick fully back I throttle up to 1,800 RPM and check the two-position propeller and then the magnetos. All is fine and so a throttle back to idle at 550 RPM confirms we are ready to go. Coolant is now 110°C!

"Lining up on the grass I gradually push the throttle forward and the aircraft starts to accelerate nicely. The first impressions are very important at this stage, I start to feel the controls come 'alive' and concentrate on whether they feel normal. Any large out of trim tendencies are better identified here on the ground rather than in the air.

"The aircraft accelerates quickly and a glance shows + 5 lbs of boost and

Above: The following marvellous sequence of air-to-air shots were captured by renowned aviation photographer John Dibbs during the aircraft's testing phase. Here, she looks glorious in her natural element. An aeroplane that former 92 Squadron pilot Geoffrey Wellum called "a lithe creature, a thing apart".

2,600 RPM. I leave the throttle there and following a few bumps of Duxford's grass we are airborne......

"I had briefed the engineers that I would not retract the undercarriage until I was happy with the engine and initial trim of the flying controls and so climbing steadily on the downwind leg I throttled back to + 2 lbs boost and selected coarse pitch on the propeller. It was like changing gear in a car from 1st to 4th! The RPM

Left: As P9374 pulls up and away from the camera, she shows off the whole of her underbelly and the distinctive black and white scheme.

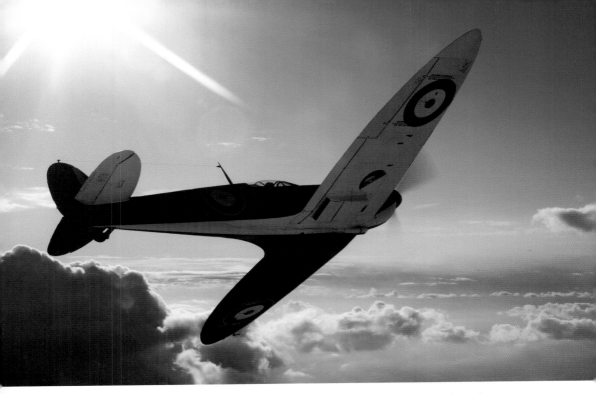

Glinting sunlight reflects the true glory of this thoroughbred of all fighter aircraft and a true masterpiece of reconstructive engineering.

dropped from 2,700 to 2,050 very smoothly and the aircraft now wanted to accelerate beyond the undercarriage speed of 160 MPH.

"The ailerons were not in trim, and so I was holding a 2-3 lb roll left tendency. However, I felt this was acceptable for now and so selected undercarriage up and started to pump the handle. There was no increase in the loads over that which I had experienced in the hangar tests, and so in good order the undercarriage retracted and the two 'up' lights showed red in the indicator.

"Closing the hood I now scanned the instruments for any sign of a problem, but all was well and so I climbed the aircraft up to 3,000 ft to carry out a stall check. In the climb I started to 'feel' the controls some more and realised that this aircraft felt really very nice. It is well balanced and apart from the out of trim aileron the controls were good. Other things started to register as 'different' from the other Spitfires. The canopy is the original flat-sided version and so I was aware of a limitation in my head movement. Basically, I kept banging my head on the canopy sides! Also the aircraft is relatively quiet in comparison to some of the later marks. The coarse pitch propeller and Siamese exhaust stacks obviously keep the noise levels lower.

"3,000ft reached and time to slow down for the stalls. The first is a 'clean' stall with undercarriage retracted and flaps up. I first trim the aircraft to fly at 105 mph and put the propeller back in to 'fine' pitch. If I require the engine to recover from the stall I will need the instant acceleration that fine pitch will give me.

"Slowing through 75 MPH the elevators give a slight buffet and the nose and left wing drop at 70 MPH. Very nice! The book figure is 73, and so this is a perfect result. Accelerating away from that, I select undercarriage down and start pumping. It takes more pumps to get the undercarriage down than it does to get it up, but still it is a painless exercise and I have not yet lost the skin off my knuckles which was apparently a common injury for many early Spitfire pilots.

Above: On final approach to Duxford after her maiden flight, P9374 curves in to land against the backdrop of a perfect and glorious summer evening.
Below: After a historic twenty-odd minutes in the air, P9374 is moments away from kissing the turf at Duxford following John Romain's first test flight.

"Slowing to below 140 MPH I select flaps down. Wow! They go down very quickly and the aircraft instantly has a nose-down pitch. Trimming back from that I note the flap extension time as 1-1 ½ seconds. The stall is again very benign and is a little under 60 MPH. The book figure is 64, so again this is a very good result.

"I have been airborne now for about 20 minutes and it is time to return for the first landing. We can then check for any leaks that may have shown up and adjust the aileron trim before testing the aircraft further.

"Flying through the airfield overhead at 500 feet I slow down to below 160 MPH and lower the undercarriage. Propeller into fine pitch, radiator door open.

Check for green down lights on the undercarriage and slow to below 140 MPH for flaps. I am ready for their rapid deployment this time and trim the aircraft

Successfully and flawlessly flown, P9374 taxies in.

to 90 MPH. Opening the hood causes a windy buffet through the cockpit, but *what* a feeling; a Spitfire I on finals to Duxford, the airfield from which the Spitfire as an RAF type first became operational in 1939. It is a lovely sunny evening and I am in an aircraft which is flying beautifully. I pinch myself and get on with the landing.

"She flares at 75 MPH and settles gently onto the grass – fantastic – no problems keeping her straight, and she slows to a walking pace for the taxi back. Puffs of oil smoke cough from the exhaust stacks as I taxi in. The warmth and smell from the engine furthering my imagination of what it must have felt like as a 20-year-old Royal Air Force pilot let loose in a Spitfire I just before the Battle of Britain. It must have been unbelievable, frightening and exhilarating all at the same time. With no slow running cut out on a Mark I it is just a case of magnetos 'off' and the engine slows and stops with a momentary 'kick' just to burn of the last bit of fuel.

"Everyone is ecstatic. And they should be. This is a fantastic restoration to be justifiably very proud of. P9374 is a truly lovely aircraft, and she flies beautifully."

Those who have been up-close and personal, seen her fly, touched her, smelt her, sat in her or even been fortunate enough to fly her will surely all agree on one thing; this Spitfire *is* P9374. Airframe, engine and soul.

APPENDIX

TECHNICAL SPECIFICATIONS OF THE SPITFIRE I AEROPLANE

The following are the leading technical specifications for the Spitfire I:

WING	Planform elliptical; section NACA 2200 series; Incidence°: at root +2, at tip – ½; Dihedral °6; Thickness % root 13.2, tip 6; aspect ratio 5.68; area sq ft net 223, gross 242; chord (geometric)6.3 (ma) 7.01
	Ailerons. Area sq ft 18.9; chord 5.54; movement° up 26, down 19,droop ⅜ in. Flaps. Area sq ft 15.6, movement° down 85.
	Wing loading lb/sq ft 36.0.
	Power loading 5.54 lb/hp.
	Wing span decreased by 3½in when Air Ministry decided that wing tips should be detachable. New tips were less pointed.
TAILPLANE	Area sq ft 31.46 with elevators; chord (ma) 4.0; incidence° root 0.
ELEVATORS	tip+- ½; dihedral 0.
TAB	Area sq ft 13.26; movement up 28, down 23.
	Area sq ft 0.38, movement° up 20 down 7.
FIN	Area sq ft 4.61.
RUDDER	Area sq ft 8.23; movement each way 28.
TAB	Area sq ft 0.35; movement° each way 12.
UNDERCARRIAGE	Wheels Dunlop AH2061; Tyres Dunlop 7-50-10.
	Oleo pressure lb/sq in 250;
	Tailwheel fixed castoring. Wheel Dunlop AH2184;
	Tyre Dunlop Ecta 3-00-4. Tyre pressure lb/sq in 60.
ENGINE	Rolls-Royce Merlin III, 12 cylinder 60° V supercharged.
	Normal bhp 950.
	Normal rpm 2600.
	Maximum bhp 1030.
	Electric starter.
	Fuel consumption 70.65 gals/hr at normal cruise power.

	12 volt system. De Havilland 3-blade, 5/29 two position propeller.
COOLANT	Treated ethylene glycol specification D.T.D. 344. Engine 4.25 gals. Radiator 4.25 gals. Piping 3.5 gals. Header tank 1.5, total 13.5 gals.
FUEL	100 octane. Capacity (fuselage) upper tank 48 gals, lower tank 37, Total 85.
OIL	5.8 gals under engine. 1.75 gals airspace in tank.
ARMOUR	73 lb including 2in thick glass windscreen and ¼ in stainless steel behind pilot seat.
ARMAMENT	Eight x Vickers Browning Mk II .303" machine guns. Firing rate 1,150 rounds per minute. Mk VII ammunition and de Wilde tracer rounds typically fitted every 54th cartridge. ('A' Wing – 4 guns each wing). Gunsight Barr and Stroud reflector type. Cine camera G42B, can be mounted on top of starboard wing. Radio. TR 9D.
AIR SYSTEM	Pneumatic system via engine-driven compressor feeding two storage cylinders holding air @ 300 lb/sq in.
HYDRAULIC SYSTEM	Hand-actuated hydraulic pump retractable undercarriage system with hydraulic reservoir behind pilot's seat.

Overleaf: Spitfire I cut away.

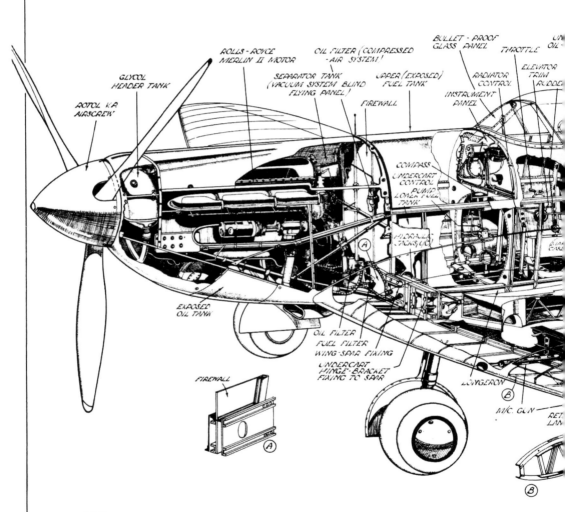

ROLLS - ROYCE
MERLIN II MOTOR

OIL FILTER (COMPRESSED
- AIR SYSTEM)

BULLET - PROOF
GLASS PANEL

THROTTLE

UN
OIL

GLYCOL
HEADER TANK

SEPARATOR TANK
(VACUUM SYSTEM BLIND
FLYING PANEL)

UPPER (EXPOSED)
FUEL TANK

RADIATOR
CONTROL

ELEVATOR
TRIM
RUDDER

ROTOL V.P.
AIRSCREW

FIREWALL

INSTRUMENT
PANEL

COMPASS

UNDERCART
CONTROL
PUMP
LOWER FUEL
TANK

HYDRAULIC
JACKS (U/C)

EXPOSED
OIL TANK

OIL FILTER

FUEL FILTER

WING-SPAR FIXING

UNDERCART
HINGE-BRACKET
FIXING TO SPAR

LONGERON

FIREWALL

M/C. GUN

RET
LAN

Ⓐ

Ⓑ

Ⓑ

DATA.

MANUFACTURERS. VICKERS - ARMSTRONG LTD. SOUTHAMPTON.
POWER PLANT. ROLLS - ROYCE MERLIN - 1,030 H.P. AT 3,000 R. AT 16,250 FT (1,250 H.P. ON 100 OCTANE)
GENERAL CONSTRUCTION. METAL, WITH STRESSED-SKIN FLUSH-RIVETED COVERING.
DIMENSIONS SPAN 36'10" LENGTH. 29'11" HEIGHT 11'5" WING AREA. 242 SQ. FT.
WEIGHT (LOADED). 5,850 lb. WING LOADING. 24.4 lb PER sq. ft. TANKAGE PETROL 85 GAL. OIL 5½ GAL.
PERFORMANCE. SPEED. 367 M.P.H AT 18,500 FT. CLIMB (MIN) 2,300 FT/MIN. DURATION 3.6 HRS (AVERAGE)

RE MARK I

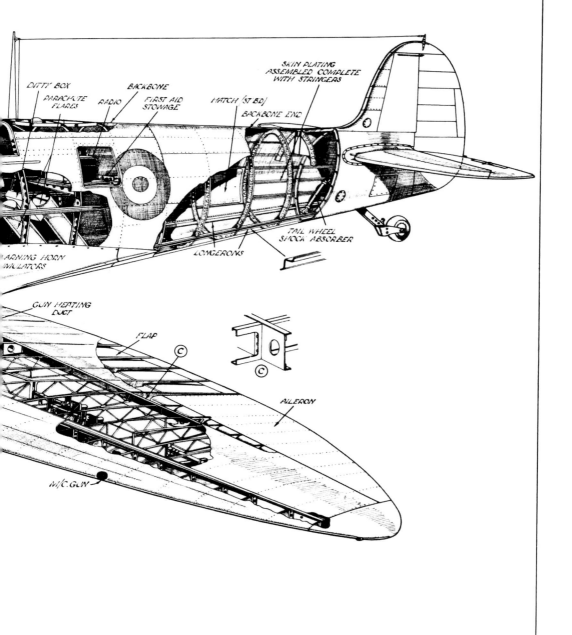

DITTY BOX
PARACHUTE FLARES
RADIO
BACKBONE
FIRST AID STOWAGE
HATCH (ST BD)
SKIN PLATING ASSEMBLED COMPLETE WITH STRINGERS
BACKBONE END
WARNING HORN INDICATORS
LONGERONS
TAIL WHEEL SHOCK ABSORBER

GUN HEATING DUCT
FLAP
Ⓒ
Ⓒ
AILERON
M/C.GUN

II APPENDIX

FOLLOW-UP TEST FLYING

In Chapter 13, John Romain described the experience of the first test flight in Spitfire P9374. Since then, the aeroplane has been subjected to a rigorous test programme of flights to satisfy the requirements of the CAA and before the aeroplane can be 'signed-off' as fully airworthy and ready for air display or other work. Here, he gives an outline of the subsequent testing of P9374 prior to its full readiness for the 2012 airshow display season, during which it will appear before the British public as the earliest mark of Spitfire flying anywhere in the world:

"After the initial first flight there followed the official air testing of the aeroplane, the nature and content of which had been discussed and agreed with the Civil Aviation Authority Flight Test Department and a format agreed upon.

"After a thorough inspection, and the first attempt at trimming the ailerons, I flew the aircraft for another twenty-minute sortie. The aileron trim had not changed very much and so I noted the need for a larger addition of cord to the trailing edge of the port aileron. (This was the standard RAF Air Publication approved 1940 method of adjusting the aileron trim.) On this sortie I also tried the first climb. The pilot's notes mentioned that the aircraft should climb at + 6 ¼ lbs boost and a RPM above 2,080. It did not, however, mention the speed. From other Spitfire flying I knew that 160-180 MPH was the norm, but at these speeds I was suffering from low RPM. What I did not want to do was 'over-boost' the engine by running a high boost setting (+ 6 ¼ lbs) with only 2,050 RPM.

"With little information in the pilot's notes I tried to obtain some original testing notes via the Boscombe Down archive. This was not forthcoming and so I tried the internet. To my surprise, a lot of original Spitfire test data is available on internet web sites, including a test carried out at Martlesham Heath on a Spitfire I with the De Havilland two-pitch propeller. These documents gave me the data needed, and also identified that the two-pitch propeller 'coarse setting' needed adjustment. It also gave me the climb speed of 185 mph, noting that this was necessary in order to obtain a favourable RPM in the climb! Finally, things were starting to make sense.

"With the propeller adjusted and another piece of cord on the aileron I took off for the third air test. The aileron was much better and the RPM in the climb sat at the correct level. This allowed further testing and so I climbed to 6,000 ft for the various stall checks and the start of the higher speed handling. Copious notes later and I landed back at Duxford, content that the aircraft was now settling down very nicely.

"The next four flights involved more climbing and stick force tests plus the start of some aerobatics. The roll rate was very similar to other Spitfire marks, but generally the aircraft was a delight to roll and to loop.

"I had attained 350 MPH in one of the dives but the aircraft had yet to be tested to the full 405 MPH as stated in the notes. A wartime figure of 450 MPH was allowed, but this meant an engine over-speed of 3,400 RPM. This speed and RPM were not necessary for our flight regime and also I wondered: how many engines survived that? An interesting letter found with the notes mentioned above, and written by Air Marshal H.C.T Dowding in early 1940, warned pilots of blowing up their engines at high RPM and boost figures.

"The dive to 405 MPH was started from 9,000 ft. As the airspeed went through 380 MPH I had 2,900 RPM and the angle of dive was fairly steep. Aileron loads were increasing, and I recalled Alex Henshaw's words that above 400 MPH they were very heavy. Just approaching this figure there was a large bang and an in-rush of air. Reducing power instantly I pulled out of the dive and regained level flight, wondering with a lot of anxiety what it was that had broken.

"It was very quickly obvious to me what the problem was, and I relaxed somewhat. The direct vision panel on the port side of the cockpit canopy had blown out and disappeared. It is simply held in place with Perspex pins and clips to allow the pilot to literally punch it out in an emergency, but the airflow in my dive had obviously been too much for it.

"So it was back to Duxford for a few circuits, both with and without flaps, and then in for repairs!

"The next day I achieved the 405 MPH dive and all went fine. However, it is a very noisy and quite startling thing to do. I can only think that combat conditions would be the only reason one would ever want to fly a Spitfire Mark I at 450 MPH. The controls would be very heavy, the engine screaming and the noise of the air-flow incredible.

"With eight flights completed the aircraft was flying very well. All the test data was written up for the report to the CAA and it was accepted the very next day – another wonderful achievement.

"The Spitfire I is certainly quite different to fly when compared to its modified and later mark companions. However, it is a delight in engineering and in pure aerodynamic terms. It is a real piece of 1938 history, and one that does everything the test flights of 1939 said it would."

What is notable throughout the efforts to get P9374 back into the air is the common thread that ran through all aspects of the project; namely, a steep learning curve. In part, this had been because of the sometimes surprising lack of detailed knowledge as to a particular instrument fit or type – or even where a certain part is actually fitted or what a stencilled servicing instruction on the wing really says. Sometimes, it has been a puzzle working out what colour a particular item might have been, or even what it looked like. A case in point was the punch-out Perspex panel referred to in

John Romain's report above, where the precise detail of how the edges of the panel were shaped and how it was held in place was simply unclear. That detail had to be deduced from period photographs and from fragmentary relic items and then manufactured on almost a best guess basis.

All of these issues had to be resolved through a combination of research and detective work, or by tracking down items that were thought to be long extinct. The meeting of all of these challenges had been made possible not only by the team directly involved with the building of P9374 but a veritable army of knowledgeable experts and enthusiasts who have all made their individual and valuable input to ensure ultimate perfection. Of course, that learning curve in relation to the build was extended as well into the aspect of test flying P9374 as we have seen from John Romain's testimony. Not since 1940 had a Spitfire to this specification been flown, and the pilot's notes and accounts and memories of those few survivors who had flown the Spitfire I were not necessarily always very helpful when it came to determining the *exact* detail and information that John needed to have in order to fly the aeroplane correctly.

Ultimately, some of that information came from the long-forgotten Martlesham Heath Spitfire I test reports. So, even the acquisition of information about the flying and handling characteristics of the Spitfire I became a challenging exercise and all part of that very steep learning curve. Whilst the uninitiated might well believe that every last detail of the Spitfire aircraft must now be well known and recorded in the twenty-first century, the experience gained when building P9374 proved that this belief was very far from the case. In almost every respect, both from building and flying P9374, those involved in the project were pretty much at the same place as those who built the first Spitfire or, to a certain extent, Mutt Summers who test flew the Spitfire on its maiden flight. John Romain's reports are certainly more than testimony enough of that inescapable fact.

Martin Overall (nearest camera) and colleague ride the tailplane back to the hangar in time-honoured fashion.

APPENDIX III

AIRCRAFT MOVEMENT CARD FOR SPITFIRE P9374 – AIR MINISTRY FORM 78

Each Royal Air Force aircraft had its individual aircraft movement card, or Air Ministry Form 78, which recorded all essential details of that individual aircraft: contract number, units to which it had been allocated, engine number, major repairs, write-off and total flying hours etc. Fortunately, the overwhelming majority of these AM Form 78s have survived for all wartime RAF aircraft and these include the form for P9374 which is reproduced here.

This is the official record of the aircraft from delivery to the RAF up until its eventual loss.

It will be seen that the aircraft was first allocated to the RAF's 24 Maintenance Unit on 15 February 1940 but almost immediately it was diverted to 9 Maintenance Unit at RAF Cosford on 28 February where it was to be prepared for delivery to 92 Squadron, newly equipping as a Spitfire unit and converting from the twin-engine Bristol Blenheim. P9374 was ultimately delivered to 92 Squadron on 6 March 1940. The aircraft is shown as having been built under contract number 980385/38 and delivered fitted with Rolls-Royce Merlin III, number 143668. The airframe is shown as ultimately accumulating thirty-two hours and five minutes flying time before being lost in a FBO (flying battle occurrence) on 24 May 1940 and struck off charge with effect from 3 June 1940. The survival of the relevant AM Form 78 in RAF records turned out to be vital in establishing the identity and history of P9374. Unfortunately, many of the AM Form 78s for surviving Spitfire aircraft were stolen from the MOD's RAF Air Historical Branch during the 1970s/1980s, but as P9374 was not then recognised as being among the known population of Spitfire 'survivors' the record card for this aeroplane was thankfully not taken. *(Note: a number of data plates found in the wreckage of P9374 from various sub-assemblies etc. all show construction dates of January 1940. Given that the delivery of the aircraft to the RAF took place in February of that year it is reasonable to assume that P9374 was on the production line and completed for delivery during January 1940.)*

Struck off charge seventy-one years earlier, P9374 is brought back to life in the workshops at Duxford during 2011.

APPENDIX IV

ENGINE HISTORY – ROLLS-ROYCE MERLIN III NUMBER 143668 (13769)

The history of the Merlin III engine fitted to Spitfire P9374 is fully recorded. It is known that it had only ever been fitted to P9374 and thus had quite limited running hours. The details are:

ROLLS-ROYCE ENGINE NUMBER:	13769
'A' ENGINE NUMBER (RAF NUMBER):	143668
FACTORY:	Derby
BUILT:	27 October 1939
TESTED:	2 November 1939
DELIVERED:	6 November 1939
DESTINATION:	14 Maintenance Unit, RAF Carlisle
ORDER NUMBER:	4790
CONTRACT NUMBER:	819222/38
FURTHER INFORMATION:	Recorded as Written-off. No further details.*
REBUILT TO AIRWORTHY:	Retro Track & Air, May 2010
GROUND TESTED ON STATIC RIG:	23 April 2010
GROUND TESTED IN P9374:	17 June 2011

*Note: Whilst the engine was to all intents and purposes written-off as a total loss by the RAF in 1940 its recovery in 1981 gave it an eventual new lease of life, with significant components from the original engine being incorporated into the rebuilt Rolls-Royce Merlin III. The rebuilt Merlin still carries its original engine number, 13769. Whilst 13769 is the engine's Rolls-Royce number, it was also allocated the Air Ministry's number of 143668, as recorded in the AM Form 78.

(The above historical information relating to the engine from P9374 was largely researched for the author by The Rolls-Royce Heritage Trust, and was gleaned from information held by them on 13 February 1981.)

V APPENDIX

MILESTONES IN THE HISTORY OF SPITFIRE P9374

Contract signed by Air Ministry for supply of 183 Spitfire I aircraft by Vickers Armstrong (Supermarine) – 1938

Delivered to 9 MU 9 March 1940

Delivered to 92 Squadron 6 March 1940

Lost on Calais beach 24 May 1940

Discovered Calais beach September 1980

Recovered to Calais Hoverport January 1981

Identified as P9374 on 17 January 1981

Transferred to Musée de l'Air, Le Bourget 19 January 1981

Purchased by Jean Frelaut circa August 1981

Purchased by Thomas Kaplan 2000

Transferred to Airframe Assemblies, Isle of Wight, 14 October 2000

Registered with CAA as G-MKIA by Simon Marsh on 16 November 2000

Transferred to storage with Charleston Aviation Services, Essex, 19 March 2002

Re-registered to Spitfire Partners LLC 3 March 2005

Wreckage transferred to Historic Flying Ltd, Duxford, August 2007

Construction of wings commenced at Duxford September 2007

Construction of fuselage commenced at Airframe Assemblies September 2007

Fuselage delivered to Duxford 8 July 2008

Re-registered to Mark One Partners LLC 13 August 2008

First post-reconstruction engine test at Duxford on 17 June 2011

First flight at Duxford 1 September 2011

APPENDIX  VI

PAINT SPECIFICATIONS

All metal camouflage surfaces were finished using Akzo-Nobel Aerodex paint in a matt finish.
All fabric-covered surfaces were finished in HMG Cellulose, and in a matt finish.

COLOURS:

Standard colours from the British Standard 381c and 4800 ranges:
Dark Green Bs381c No. 241. Top camouflage
Dark Earth Bs381c No. 350. Top camouflage
Night Bs381c No. 642. Lower port wing, all stencils, propeller spinner
Midnight Bs4800 20 C 40. Roundels
Golden Yellow Bs381c No. 356. Roundels
Aircraft Grey/Green Bs381c No. 283. Interior components
Banana Bs4800 10 D 43. Gas patch
White (No British Standard number allocated). Lower starboard wing, stencils, roundels, fin flash stripe

Wartime colours not incorporated into the British Standard range but matched to original wartime colour chips:
Red (*not* postwar bright red, Post Office or roundel red). Roundels, fin flash stripe, gas patch edge strip.

OTHER COLOURS:

Matt silver for interior of fuselage
Black for instrument panel, various components
Satin silver for some internal components

Note: the paint colours used are the modern equivalents of the wartime RAF colours that were given a D.T.D. classification, with the initials standing for the Air Ministry Directorate of Technical Development.

VII APPENDIX

ALL KNOWN RECORDED FLIGHTS BY P9374

Due to the fact that the operations record book for 92 Squadron does not start to record details of individual flights by pilots and aircraft until 9 May 1940, it has not proved possible to detail every flight carried out by P9374 whilst the aircraft was with 92 Squadron. However, the known and recorded flights are detailed as under:

9 May 1940	Sgt Fokes	14.45 – 15.40	To Northolt from Croydon
10 May 1940	Fg/Off Cazenove	10.15 – 11.05	Sector recce
10 May 1940	Plt/Off Bryson	11.35 – 12.15	Practice attacks
10 May 1940	Fg/Off Cazenove	14.35 – 15.00	Formation flying
14 May 1940	Fg/Off Cazenove	09.30 – 10.20	Practice attacks
14 May 1940	Fg/Off Cazenove	21.20 – 21.30	Dusk landings
14 May 1940	Fg/Off Cazenove	22.30 – 23.35	Sector recce
15 May 1940	Plt/Off Bryson	00.05 – 00.45	Sector recce
15 May 1940	Fg/Off Cazenove	15.25 – 16.00	Aerobatics
15 May 1940	Sgt Barraclough	18.50 – 19.20	Night flying test
16 May 1940	Plt/Off Saunders	10.50 – 11.45	Circuits and landings
17 May 1940	Sgt Eyles	20.25 – 20.45	Night flying test
18 May 1940	Plt/Off Bryson	22.05 – 23.00	Night flying recce
19 May 1940	Flt/Lt Green	07.05 – 07.25	To Hendon
19 May 1940	Plt/Off Saunders	08.40 – 08.50	From Hendon
19 May 1940	Sgt Barraclough	20.10 – 20.40	Night flying test
20 May 1940	Fg/Off Cazenove	11.45 – 12.50	Formation and target
20 May 1940	Fg/Off Cazenove	14.55 – 15.45	Formation and attacks
23 May 1940	Plt/Off Williams	17.20 – 19.05	Patrol Calais/Boulogne/Dunkirk
24 May 1940	Fg/Off Cazenove	08.05 – c.09.00	Patrol Calais/Boulogne/Dunkirk

NOTE: Although flying details and the hours flown for P9374 are recorded (as above) in the 92 Squadron operations record book from 9 May through to 24 May 1940, the details for 21 and 22 May are missing from the squadron records although it may reasonably be assumed that P9374 flew at some time during that forty-eight hour period. Indeed, and although P9374 was washed out of its intended escort flight to France on 19 May, the possibility cannot be excluded that P9374 took part in the escort flights to and from France which the squadron conducted over 21 and 22 May in order to protect Prime Minister Winston Churchill.

The above flying hours total fourteen hours and twenty minutes, leaving a shortfall of seventeen hours forty-five minutes in respect of the officially recorded flying hours of P9374 up until its loss on 24 May 1940. Of this, it could be reckoned that some two hours would have been spent in the air on delivery from Southampton to RAF Cosford, and from Cosford to 92 Squadron's Croydon base. The detail of the remaining flight time for P9374 remains unknown at time of publication.

VIII APPENDIX

ALL KNOWN RECORDED FLIGHTS BY
FG OFF PETER CAZENOVE WITH 92 SQUADRON

Date	Aircraft	Time	Activity
9 May 1940	N3194	14.45 – 15.40	Northolt to Croydon
10 May 1940	P9374	10.15 – 11.05	Sector recce
10 May 1940	P9374	14.35 – 15.00	Formation flying
14 May 1940	P9374	09.30 – 10.20	Practice attacks
14 May 1940	P9374	21.20 – 21.30	Dusk landings
14 May 1940	P9374	22.30 – 23.35	Sector recce
15 May 1940	P9374	15.25 – 16.00	Aerobatics
15 May 1940	P9373	17.55 – 18.45	Formation & attacks
16 May 1940	N3287	14.55 – 15.50	Camera gun attacks
16 May 1940	N3194	20.20 – 20.35	Night flying
16 May 1940	N3194	22.15 – 23.15	Sector recce
17 May 1940	N3248	14.15 – 14.45	Camera gun attacks
17 May 1940	N3248	15.40 – 16.05	Camera gun attacks
18 May 1940	N3194	09.25 – 10.25	Air firing
19 May 1940	P9373	16.40 – 17.30	Sector recce
20 May 1940	P9374	11.45 – 12.50	Formation & target
20 May 1940	P9374	14.55 – 15.45	Formation & attacks
24 May 1940	P9374	08.05 – c.09.00	Operational patrol

The above is merely a snap-shot of Fg Off Peter Cazenove's flying with 92 Squadron in May 1940 and presents, unfortunately, a far from complete picture based upon the only evidence available to us. It will be seen that we have a record of Cazenove flying P9374 on seven different occasions and if it was not allocated as 'his' Spitfire then it certainly seems to be the case that it was the aircraft he preferred to fly whenever possible. Unfortunately, Cazenove's flying logbook does not survive.

APPENDIX IX

ALL KNOWN VICTORY CLAIMS BY 92 SQUADRON
23 AND 24 MAY 1940

23 MAY 1940

Flt Lt R R S Tuck	Me 109 (destroyed)	Boulogne/Dunkirk
Plt Off C P Green	Me 109 (unconfirmed)	Boulogne/Dunkirk
Plt Off J S Bryson	Me 109 (destroyed)	Boulogne/Dunkirk
Plt Off A C Bartley	Me 109 (unconfirmed)	Boulogne/Dunkirk
Plt Off H D Edwards	Me 109 (unconfirmed)	Boulogne/Dunkirk
Plt Off H D Edwards	Me 109 (unconfirmed)	Boulogne/Dunkirk
Sgt S M Barraclough	Me 109 (unconfirmed)	Boulogne/Dunkirk

All above claims were made during the morning action.

Flt Lt R R S Tuck	Me 110 (destroyed)	Off Boulogne
Flt Lt R R S Tuck	Me 110 (unconfirmed)	Off Boulogne
Sqn Ldr R Bushell	Me 110 (damaged)	Off Boulogne
Sqn Ldr R Bushell	Me 110 (damaged)	Off Boulogne
Plt Off A C Bartley	Me 110 (destroyed)	Off Boulogne
Plt Off A C Bartley	Me 110 (destroyed)	Off Boulogne
Plt Off R H Holland	Me 110 (destroyed)	Off Boulogne
Plt Off R H Holland	Me 110 (destroyed)	Off Boulogne
Plt Off R H Holland	Ju 88 (destroyed)	Off Boulogne
Sgt R E Havercroft	Me 110 (unconfirmed)	Off Boulogne
Plt Off A R Wright	Me 110 (destroyed)	Off Boulogne
Plt Off H D Edwards	Me 110 (unconfirmed)	Off Boulogne
Plt Off D G Williams	Me 110 (unconfirmed)	Off Boulogne

| Sgt S M Barraclough | Me 110 (unconfirmed) | Off Boulogne |
| Plt Off H D Edwards | Me 110 (destroyed) | Off Boulogne |

All of the above claims were made during the afternoon action. A 'general combat report' for that action was filed by 92 Squadron:

> "No 92 Squadron while on patrol Boulogne-Calais-Dunkirk from 17.20 to 19.20 hours sighted a large formation of enemy aircraft over Calais at 8 – 10,000 feet. About thirty enemy aircraft, mostly Me 110s, were ahead and behind them a further group of fifteen to twenty aircraft of types not identified, but including Ju 87s and Ju 88s. Some aircraft started dived bombing attacks on Boulogne harbour, with others circling in the vicinity. No 92 Squadron at 4,000 feet* climbed to engage and a series of dogfights ensued, mainly with Me 110s. Blue 1 states that he saw some Hurricanes already engaging the enemy, but as the sky was so full of aircraft a clear statement of the situation is impossible. As a result of the dogfight seven Me 110s were definitely shot down and five Me 110s and two Ju 88s were probably shot down. Most of our aircraft were hit many times.
>
> Pilots state that the Me 110's evasive tactics are a steep turn towards the Spitfire's tail, to enable the rear gunner to open fire. [*sic*]
>
> About twenty Me 110s were seen flying in line astern in a tight circle around the bombers, which was very difficult to attack. The Me 110 is not so fast as the Spitfire on the level, but very good in a fast turn and a steep dive, though the Spitfire can hold it on a turn. They appear to use the stall turn a great deal.
>
> Enemy camouflage standard.

* NOTE: This is the first indication of the altitude 92 Squadron were operating at that day, and at 4,000 feet it is perhaps a surprisingly low combat patrol altitude. In the transit from Hornchurch to the patrol line the squadron should, perhaps, have had ample time to climb to greater altitude in order to achieve a height advantage. At this relatively low altitude the squadron were potentially placing themselves at a distinct height disadvantage, and were thus setting themselves up for a perfect 'bounce' by German fighters who would inevitably be operating at a much higher altitude. The squadron was also operating at an altitude where they would most likely have to climb in order to attack any enemy bombers. That said, it perhaps remains possible that the lack of constant-speed propeller units to the squadron Spitfires at this time may have had some bearing on this lack of altitude by the time the squadron had reached the French coast. The rate of climb was significantly improved when the c/s units were fitted to Spitfires from 25 June 1940 onwards, a conversion that also gave an additional 7,000 feet of altitude. More likely, though, it was more to do with a leisurely climb and transit across the Channel by a combat-inexperienced squadron for whom the possibility of engagement with the enemy probably seemed rather unreal. As we now know, the squadron had had a rude awakening that morning.

During the first two minutes of the combat a continuous transmission in German was heard on the R/T."

24 MAY 1940

Flt Lt R R S Tuck	Do 17 (destroyed)	Calais/Dunkirk
Flt Lt R R S Tuck	Do 17 (destroyed)	Calais/Dunkirk
Plt Off A C Bartley	Me 110 (destroyed)	Calais/Dunkirk
Plt Off H D Edwards	Me 110 (destroyed)	Calais/Dunkirk
Plt Off R H Holland	Do 17 (unconfirmed)	Calais/Dunkirk
Sgt R E Havercroft	Do 17 (unconfirmed)	Calais/Dunkirk

(All of the above claims were made during the early evening action.)

It is important to note that these are claims made by squadron pilots and 'allowed' at the time on the basis stated in brackets. Subsequent research indicates there was clearly a large degree of over-claiming. One further claim for a Dornier 17 (confirmed) was made by an unknown pilot on 25 May before the squadron was withdrawn to RAF Duxford.

X APPENDIX

TWO FOR THE FUTURE?

SPITFIRE P9372

On 5 March 1940 the 92 Squadron operations record book recorded that the squadron had been allocated twenty-one Spitfires; eight from 9 MU, RAF Cosford, and thirteen from 27 MU, RAF Shawbury. Along with P9374, a consecutive 'batch' of Spitfire I aircraft were delivered to 92 Squadron on or around 6 March 1940 from this Air Ministry allocation held at the two maintenance units.

The Aircraft Movement Card (Air Ministry Form 78) for Spitfire P9372.

These aircraft included P9372 and P9373, but P9367, P9368, P9369, P9370 and P9371 were also in the same serial batch and were delivered at around the same time from to 92 Squadron.

Almost certainly, P9372 was one of seven Spitfires collected from Cosford on 6 March, and ferried down to RAF Croydon by a disparate group of pilots; Sqn Ldr Roger Bushell (92 Sqn), Flt Lt Adrian Boyd (145 Sqn), Sqn Ldr Pat Gifford (3 Squadron) and three other un-named pilots of 65 Squadron. The reason a decision was made to deliver Spitfires to the squadron by non-squadron pilots was simply because these selected 'ferry pilots' all had extensive Spitfire experience and it had only been the previous day, 5 March, that squadron pilots had first been introduced to a Spitfire (of 65 Squadron) in a hangar at RAF Northolt. Gradually, the pilots would be weaned onto the fighter, a process that began on 8 March with lectures on Spitfire flying by Flt Lt Boyd of 145 Squadron followed by a visit from a Mr Rose of Vickers Supermarine to talk about technical aspects of the aircraft.

As with P9374, the flying history of P9372 is relatively unknown and it was not until 9 May 1940 when detailed entries of flights were written up for individual aircraft in the squadron operations record book. In the case of P9372, however, no record of the aeroplane having flown with the squadron is recorded until 23 May when it takes part in the fateful first operation of 92 Squadron and is flown by Sgt Barraclough. The next day, when P9374 is lost with Fg Off Cazenove, P9372 is flown by Plt Off Tony Bartley.

Although we know that P9372 had been delivered to 92 Squadron on 6 March it is possible that this Spitfire had been a reserve aeroplane on the squadron and was only flown operationally for the first time on 23 May. On the other hand, it might simply have been the case that P9372 was undergoing maintenance or repairs between 6 and 23 May. What is certainly true, however, is that P9372 was definitely undergoing repairs after 24 May! On that particular day, and with the aircraft being flown by Bartley in what was the penultimate operation of the period for 92 Squadron along the Dunkirk, Calais and Boulogne coastline, the aircraft was badly damaged by machine-gun fire. Graphic accounts of that action are given in Larry Forrester's *Fly For Your Life* (the biography of Stanford Tuck) and *Smoke Trails in the Sky* (the autobiography of Tony Bartley). First, the Stanford Tuck account:

> "Tuck joined up with another Spitfire midway over the Channel. It was Tony Bartley, who greeted him flippantly.
>
> "'I say, old boy, what you got there – a piece of lace?'
>
> "'You've collected a few holes yourself,' Tuck told him. 'Better stick your arms out and start flapping!'
>
> "Each made a careful survey of the other's aircraft and they exchanged details of visible damage. Then they arranged a bet – on which of them had the most holes. The loser would stand the winner free beer all that evening.
>
> "As they crossed the south coast Tuck's engine began to rasp and clank and falter. The cooling system was useless, and rising pressures and temperatures were bending the dial needles against the stop-studs. He throttled back a little and nursed her along, losing precious height but keeping the prop turning. Tony opened his canopy, shut off his engine and glided in close, listening. 'You'd better go in first, old boy,' he offered. 'You sound like a tinker's barrow.' Tuck accepted, gratefully."

Spitfire I, P9372, in flight whilst in service with 92 Squadron, May 1940.

Notwithstanding the damage to Tuck's Spitfire (N3192) he was flying in the same aircraft again just a little over twenty-four hours later, but the damage to Bartley's aircraft (P9372) was rather more significant. Later, Bartley was able to claim his free beer after a hole count by the fitters and riggers had revealed that P9372 had taken twelve wounds more than Tuck's. Bartley takes up the story in his own words, which largely mirror Tuck's version of events:

> "As I was racing back across the Channel, another Spitfire flew up beside me, and the pilot pulled back the hood and started pointing at my aircraft. Then, Bob Tuck came on the intercom and chortled, 'You look like a sieve, chum.' I scanned his fuselage and answered back, 'Just wait until you get a look at your crate. We escorted each other, more slowly, back to Hornchurch."

Whilst both pilots broadly agreed on events as they returned from the battle that had left Peter Cazenove on Calais beach, each of them mistakenly identify that this took place during the air battles of 23 May. In fact, it did not occur that day and there can be no doubting that these events happened on 24 May. Either way, P9372 would be out of the fray for some months and was transferred to the care of an RAF maintenance unit on 1 June 1940. Thereafter, she underwent extensive repairs before being transferred to 6 Maintenance Unit, RAF Shawbury on 21 July, preparatory to being returned back to 92 Squadron. The squadron eventually took her back on 27 July 1940, just as the Battle of Britain was getting into its stride. At this point, it is necessary to fast-forward events to 9 September 1940.

Still with 92 Squadron, P9372 was being flown on 9 September by twenty-year-old Plt Off W C 'Bill' Watling. Airborne at around 16.30 hours from Biggin Hill, the squadron was ordered to patrol Canterbury with 41 Squadron's Spitfire up to cover nearby Maidstone. The official Air Ministry (RAF Historical Branch) narrative later told how 41 and 92 Squadrons "took the first shock" of the incoming German attack over Kent and Sussex, a force comprising "a large formation of bombers escorted by Me 109s". Somewhere, and somehow, in the whirling mêlée that followed Watling's Spitfire was hit and set ablaze at around 20,000 ft over the Sussex harbour town of Rye. Watling was faced with no other option than to bale out, and leaving P9372 to its fate he drifted down to land in the sea off Winchelsea Beach from where he was eventually rescued suffering from bad burns on his face and hands. Meanwhile, the Spitfire had dived vertically under power into the soft marshland at East Guldeford, just to the north-east of Rye. On impact it had simply sliced into the soft soil and vanished, leaving little trace apart from the disturbed soil of a shallow crater and two slash marks either side where the port and starboard wings had cut through the turf as neatly and cleanly as a hot knife through butter. As a total loss, and beyond any practical recovery in 1940, Spitfire P9372 was simply abandoned by the RAF and forgotten until the 1980s when enthusiast Malcolm Pettit located and excavated the deep wreckage in the late 1980s. What he found was quite astonishing in terms of buried aircraft wrecks dating from the 1940 period. And there was also some fascinating evidence of P9372's earlier battles over Calais on 24 May, too.

Amongst the tangled pieces of wreckage there emerged large sections of fuselage and tail, still bearing original camouflage paint and the battle-damage repair patches that had been fixed over P9372's earlier wounds and scars. More intriguingly, perhaps, and certainly leaving no doubt as to the identity of the aircraft, were two large data plates found in the wreckage that were stamped P9372. These advised that in removing and then replacing the wings during its post 24 May 1940 repairs, over-sized bolts had been fitted into the appropriately reamed-out holes. To leave no doubt as to the identity of the aeroplane and its matching wings in the event that they should ever be needed to be detached again, these plates were applied to record the bolt size details and the number of the aeroplane; P9372. They were certainly convenient 'identifiers' for the wreckage, although, as we have already seen in earlier chapters, there had been nothing so stridently obvious by which to identify P9374.

The wreckage of P9372 went into Malcolm's private museum collection until 2011 when he sold it and its all-important identity to Peter Monk. At the time of writing it awaits attention on Peter Monk's Spitfire 'production line' which is gradually seeing more than one early mark of Spitfire back to life. Many original components of P9372 will be incorporated into the rebuild and although some will find it quite surprising that an aircraft that has driven itself deep into the ground can be resurrected to the point that it will fly again, it must be borne in mind that it is the iden-

The wing over-size bolt data plate recovered from the wreckage of Spitfire P9372 was actually vital in establishing the constructor's unique fuselage number for P9374. The constructor's number for P9374 had previously been unknown, but the discovery of this data plate in P9372 (giving the constructor's number 6S 30583 enabled the constructor's number of P9373 to be established as 6S 30564 and, hence, P9374 deduced as 6S 30565. Once again, the value of aviation archaeology was illustrated.

tity and a robust and continuing provenance of the airframe, as well as the inclusion of significant original parts, that drives forward such projects and makes them possible. After all, and whilst P9374 emerged from the sand still looking pretty much like a Spitfire, the starting point when it came to the reconstruction project was probably not too far away from Peter Monk's with P9372. Either way, P9372 is to be registered with the UK's Civil Aviation Authority but is not currently on the civil aircraft register. However, it is likely to be only a few years until P9372 is seen in the skies again, and perhaps airborne alongside P9374. Both aeroplanes were theoretically but one fuselage removed from each other on the 1940 production line, and both flew together in 92 Squadron. Incredibly, there is yet another Spitfire from this same aircraft serial number run, and from 92 Squadron, that also waits for eventual re-build and return to flight. It was also very much involved in the frenetic activities of 92 Squadron during May 1940. That aircraft is Spitfire P9373.

SPITFIRE P9373

Allocated to 92 Squadron in March 1940 in the batch P9367 – P9374, this aircraft was one of those lost during the second and most costly operational flight flown by the squadron on 23 May 1940. This was the afternoon 'op' when Flt Lt Green was severely wounded, Sqn Ldr Bushell and Fg Off Gillies taken prisoner of war and Sgt Klipsch, pilot of P9373, shot down and killed. Klipsch,

The Aircraft Movement Card (Air Ministry Form 78) for Spitfire P9373.

following his pilot training, was promoted to Sgt on 27 January 1940 and he was posted to 92 Squadron on 10 February 1940. *(Note: although we do not have the date or details we do know that Paul Klipsch 'buzzed' his mother's house at Watford in a Spitfire during his time with 92 Squadron – a not untypical action of an exuberant young fighter pilot of the period.)*

The action on 23 May 1940, of course, was that described in the 92 Squadron operations record book as 'a glorious' day and in which a massive over-claiming of victories was made by squadron pilots. On the ground that day, German troops, artillery and mechanised armour were encircling a beleaguered Calais and moving steadily on the port throughout the day. Just a few miles away, the Germans were moving towards the village of Wierre Effroy when at around 7pm (local time) a Spitfire dived vertically out of the air battle above and crashed into farmland there deeply burying itself on impact. Local brothers, Auguste and René Mierlot, recalled the drama of those events and how the local undertaker buried the pilot's body in the village churchyard and risked the anger of German troops who entered the village not long after the crash. It had been Sgt Paul Klipsch. Shot down during a furious engagement with the Messerschmitt 110 of II Gruppe ZG 76, most likely falling to the guns of Fw Hahn of the 6th Staffel. In Larry Forrester's biography of Bob Stanford Tuck there is an account of the loss of Sgt Klipsch, although inexplicably he refers to him as 'Flt Sgt Wooder'. Quite possibly this was simply to disguise the pilot's Germanic sounding surname in this 1956 publication, such were the sensitivities relating to these things even some ten years after the end of the war. It is difficult, really, to find any other explanation. However, the account of P9373 going down is worthy of inclusion here:

> "To his right he saw Flt Sgt Wooder's *[sic]* Spitfire a blazing brand. Under the Perspex of its canopy there was not the vaguest sign of a human figure, only a seething mass of yellow flame. It was like looking through the peep-hole of a blast furnace."

Postwar, and Klipsch's grave remained in situ at Wierre Effroy where it was marked by a Commonwealth War Graves Commission headstone and during the 1980s this attracted the interest and attention of a British researcher and enthusiast, the late Alan Brown. Asking around the village about the story of the RAF pilot, Al was soon directed to a farmer's field in a lane not far distant from the church and here he immediately started to find identifiable Spitfire debris. When, in the late 1990s, enthusiast and present-day Spitfire builder Steve Vizard was asked to find an interesting aircraft crash site to excavate for the popular Channel 4 TV series Time Team, he suggested the Wierre Effroy location and on 1 June 1999 the aircraft was recovered for the cameras in a project masterminded by Wildfire TV, the Time Team production company.

The significant portion of the aircraft fuselage and engine was recovered, from tail wheel to propeller hub. Compressed into a column of aluminium wreckage was also the remnants of the cockpit wreckage and it was here that a particularly poignant find was made in the form of an RAF flying map on which was written the name 'Sqn Ldr Roger Bushell'. Evidence of Fw Hahn's cannon and bullet strikes were still visible in the recovered wreckage. In that same action, and only a few miles distant from the crash site of P9373, Bushell had been shot down and captured. Clearly, he had flown P9373 at some stage and his maps had been left in the cockpit and although this Spitfire was regularly the one flown by Klipsch, the record shows that Bushell flew it at least once, on 9 May 1940. So, too, had P9373 been flown by Fg Off Peter Cazenove; most recently

The crash site of Spitfire P9373 was excavated at Wierre Effroy in the Pas de Calais during June 1999 for the popular British TV archaeology series Time Team. Here, the compacted wreckage of the fuselage is exposed with the tail-wheel tyre just visible.

on a sector reconnaissance on 19 May. Just like the wreckage of P9374 many years previously, positive identification of the aircraft was provided by the serial number stencilled clearly in bright red paint on the .303 Browning machine-gun ejector chutes (see photo on page 18).

The recovered wreckage was returned to the UK, initially intended for display at the Front Line Aviation museum on the Isle of Wight. However, when that museum folded the wreckage was in storage with Steve Vizard who later sold it to the owners of P9374, Mark One Partners. P9373 was placed on the civil aircraft register on 10 September 2008 as G-CFGN. At the time of writing, the wreckage remains in storage at Duxford pending a decision on the commencement of yet another Spitfire I reconstruction project that will hopefully see P9373 returned to flight.

NOTE: Whilst the wreck of P9373 awaits further attention and re-construction, Mark One Partners have since commissioned work to begin on another Spitfire I project recovered from France, this time N3200 of 19 Squadron. Again, N3200 had been force-landed on the beach, this time near Sangatte. Its pilot, the 19 Squadron commanding officer, Sqn Ldr G D Stephenson, had been taken prisoner of war after being shot down on 26 May 1940 and had ended his years of incarceration at the infamous Colditz castle. The wreck was recovered in 1986 and initially displayed at Fortresse de Mimoyecques from where it was acquired by Thomas Kaplan in December 2000. Work on this aircraft is on-going at the time of going to print. The aircraft was registered with the Civil Aviation Authority as G-CFGJ on 11 August 2008.

This superb photograph of a Spitfire I at the gun butts might otherwise have remained an anonymous aircraft were it not for the constructor's number 6S 30564 stencilled on the starboard side of the fuselage behind the cockpit. This allows us to identify the aircraft as P9373 following the discovery of the constructor's number for P9372.

Thus, in future years it is possible that P9372, P9373 and P9374 will all be in the air together. Just as they once were in 1940.

SPITFIRE P9374 DEDICATED WEBSITE

The owners of P9374 have established an official website dedicated to the aircraft with input from the author. It may be found here:

http://www.markonepartners.co.uk/

END NOTE: As far as the author has been able to ascertain, P9374 had never visited RAF Duxford in her previous life. It is interesting to reflect, though, that had she survived the fighting over the French beaches on 24 May 1940 then P9374 would have landed on the Duxford grass during the afternoon of 25 May when what remained of the squadron was withdrawn there. But then of course, had that happened, P9374 would surely not have been resurrected to her full glory as she was in 2011!

SELECTED BIBLIOGRAPHY

Bartley, Anthony DFC	*Smoke Trails in the Sky*	(William Kimber & Co 1984)
Brickhill, Paul	*Reach For The Sky*	(Collins 1954)
Churchill, Winston S.	*Their Finest Hour – The Second World War, Volume Two*	(Cassell & Co 1949)
Cornwell, Peter D.	*The Battle of France – Then and Now*	(After the Battle 2007)
Foreman, John	*RAF Fighter Command Victory Claims 1939-1940*	(Red Kite 2003)
Forrester, Larry	*Fly for Your Life*	(Frederick Muller 1956)
Halley, James J.	*The Squadrons Of The Royal Air Force & Commonwealth 1918-1988*	(Air Britain 1988)
Kingcome, Brian	*A Willingness to Die*	(Tempus Inc 1999)
Macmillan, Capt Norman	*The Royal Air Force in The World War Volume Two, 1940-1941*	(Harrap 1944)
Morgan, Eric B. & Shacklady, Edward	*Spitfire – The History*	(Key Publishing 1987)
Ponting, Clive	*Churchill*	(Sinclair-Stevenson 1994)
———	*Royal Air Force Aircraft Serials K1000 – K9999*	(Air Britain 1976)
	L1000 – L9999	
	N1000 – N9999	
	P1000 – R9999	
Price, Alfred	*The Spitfire Story*	(Cassell & Co 1995)

Price, Dr. Alfred & Blackah, Paul	*Supermarine Spitfire – Owners' Workshop Manual*	(Haynes 2007)
Price, Alfred	*Spitfire – A Complete Fighting History*	(The Promotional Reprint Company 1991)
Ramsey, Winston G.	*The Battle of Britain – Then and Now*	(After the Battle 1980)
Riley, Gordon Arnold, Peter & Trant, Graham	*Spitfire Survivors – Then and Now*	(A-Eleven Publications 2010)
Robertson, Bruce	*Spitfire – The Story of a Famous Fighter*	(Harleyford 1960)
Robinson, Michael	*Best of the Few – 92 Squadron 1939-40*	(Michael Robinson 2001)
Shores, Christopher and Williams, Clive	*Aces High*	(Grub Street 1994)
Smith, Richard	*Hornchurch Scramble*	(Grub Street 2000)
Tuck, Robert R S	*Flying Log Book (facsimilie)*	(After The Battle 1996)
Vacher, Peter	*Hurricane R4118*	(Grub Street 2005)
Wellum, Geoffrey	*First Light*	(Viking 2002)
Wynn, Kenneth G.	*Men of the Battle of Britain*	(CCB Associates 1999)
(No author)	*The Air Force List*	HMSO 1940

INDEX

PEOPLE

Arnold, Peter 27
Aschan, Jean 35
Barbas, M 5
Black, Guy 83, 91
Brown, Alan 159
Cazenove, Edna 21, 29, 31, 34, 48, 49
Charleston, Craig 60
Churchill, Winston 46, 48, 49, 50, 56, 149
Daladier, Eduoard 46
Deal, Lewis 81
Ditheridge, Tony 71
Duquenoy, M 5
Eden, Anthony 55
Forrester, Larry 21, 54, 155, 159
Frelaut, Jean 57, 59
Frewen, Adml Sir John 29
Frewen, Lady Jane 29
Gamelin, M Gen 46
Henocq, Martin 88
Henshaw, Alex 141
Kaplan, Thomas 3, 59, 60, 61, 69, 161
Kindblom, Sven 102
Louf, Jean 4, 5, 6, 7, 8, 12, 13, 14, 15, 17, 19,
    21
Loweth, John 121
Marsh, Simon 59, 60
Mierlot, Auguste 159
Mierlot, René 159
Mitchell, R J 11
Monk, Peter 53, 59, 157, 158

Norfolk, Chris 88, 114, 120, 121
Overall, Martin 68, 85, 90
Parr, Martin 88
Pettit, Malcolm 157
Pope, Col 113, 114, 119, 120, 123
Reynaud, M 46
Rickards, Steve 66, 67, 68, 82
Rivers, Paul 88
Romain, John 100, 114, 126, 140, 142
Summers, J 'Mutt' 142
Vizard, Steve 60, 76, 78, 83, 84, 85, 160
Watts, Peter 98, 99, 100, 101, 107, 112
Watts, Stuart 106, 107, 112

MILITARY PERSONNEL

Ahrens, W Ofw 48
Aubert, R D Plt Off 26
Bader, D R S Flt Lt 109
Barraclough, S M Sgt 38, 39, 155
Bartley, A C Flt Lt 22, 23, 24, 52, 155
Boyd, A H Flt Lt 155
Branch, G R Fg Off 36
Browning, General 38
Bryson, J S Plt Off 39, 40, 48
Bushell, R J Sqn Ldr 32, 34, 36, 49, 50, 155,
    158
Cazenove, P F Fg Off 20, 21, 23, 24, 25, 29,
    30, 32, 33, 40, 44, 45, 46, 52, 55, 155
Coningham, A Air Marshal, Sir 38
Dowding, ACM H C T 20, 45, 141

Drummond, J F Fg Off 44

Eyles, P R Sgt 40

Fiske, W M L Plt Off 36

Fokes, R H Sgt 42, 43, 47

Gifford, P Sqn Ldr 155

Gillies, J A Fg Off 37, 50, 52, 158

Green, C P Flt Lt 36, 37, 38, 47, 50, 55, 158

Hahn, Fw 159

Holland, R H Plt Off 52

Klipsch, P Sgt 50, 51, 158, 159

Langenburg, Fw 51

Learmond, P A G Plt Off 25, 26, 37, 48, 49, 52, 55

Montgomery, B Field Marshal 38

Potzsch, A Fw 48

Sanders, P J Sqn Ldr 55

Saunders, C H Plt Off 41, 42, 47

Stanford Tuck, R R Flt Lt 21, 22, 28, 48, 52, 155, 156

Stephenson, G D Sqn Ldr 161

Treacy, W P F Flt Lt 33

Wade, T S Plt Off 48

Watling, W C Plt Off 41

Wellum, G H A Plt Off 48, 52, 65

Williams, D G Plt Off 43, 44, 52

PLACE NAMES

Abbeville 49

Aberdeen 21

Angmering-on-Sea 35

Barth 34

Boulogne 37, 51

Brookland 41

Broxbourne 29

Calais 2, 4, 5, 6, 7, 8, 13, 14, 15, 19, 20, 24, 26, 33, 37, 46, 47

Canterbury 156

Cap Gris Nez 37

Carlisle 14

Chateau-de-St.Loup 43

Chelsea 30, 35

Chichester 38

Croydon 41

Derby 14, 41

Desvres 33, 34

Dover 37

Dungeness 40

Dunkirk 37, 40, 42, 48

Dursley 61, 95

Dutton Homestall 37

East Guldeford 157

English Channel 14

Etaples 49

Eton 29, 31

Falmouth 40

Goodwin Sands 8

Gravelines 48

Hampton 42

Hatfield 30, 31

Hendon 19

Hove 44

Kew 28

Knightsbridge 30

Le Bourget 2, 14, 57

Les Salines 26

Little Marcle 43

Littlehampton 21

Livingstone 38

Llandaff 40

London 46

Lynton 40

Maidstone 156

Montreal 39

Nairobi 35

North Weald Bassett 40

Ontario 39

Papua New Guinea 4

Paris 15

Pevensey Bay 9

Phare-de-Walde 5, 24, 27

Pietermaritzburg 36

Portslade 44

Prenzlau 34

Quai d'Orsay 46

Quebec 39

Ramsgate 15
Rawarsh 41
Roehampton 37
Runnymede 25, 40
Rye 41, 157
Sagan 34
Salisbury 44
Sandown Airport 60
Seaford 29
Sergoit 35
Shoreham 38, 73
Shorncliffe 37
South Lambeth 79
St Moritz 36
Terclinthun 26
Trier 34
Truro 40
Waldam 5, 10
Westmount 39
Wierre Effroy 50, 159
Winchelsea Beach 157

Hawkinge 37
Hendon 36, 38, 47
Henlow 40
Heston 42
Hornchurch 23, 37, 52, 156
Kemble 100
Kenley 30
Lentini East 41
Maison Blanche 38
Manston 41
Martlesham Heath 140
Netheravon 39
North Weald 117
Northolt 33, 36, 47, 48, 155
Old Sarum 30
Shawbury 40, 43, 156
Snailwell 43
St. Inglevert 51
Syerston 39
Tangmere 36, 40
Thorney Island 30
West Malling 37
Yatesbury 42

RAF AIRFIELDS AND ESTABLISH-MENTS

Abu Seir 41
Bassingbourn 39
Bexhill-on-Sea 42
Biggin Hill 37, 156
Boscombe Down 140
Church Fenton 38
Cosford 28, 149, 155
Croydon 32, 33, 46, 149, 155
Drem 40
Duxford 40, 55, 60, 61, 79, 81, 141, 160
Eastchurch 41
Fairwood Common 38
Gatwick 32
Gravesend 41
Habbaniya 40
Halton 39
Hanworth 42, 43
Hawarden 39

RAF AND LUFTWAFFE UNITS AND OTHER MILITARY FORMATIONS

CFE 38
CFS 39
Rifle Brigade 49
Queen Victoria Rifles 49
1./JG 27 48
5./ZG 76 51, 52
1 EFTS 31
1 ITW 42
3 Sqn 155
5 FTS 42
7 OTU 39
9 MU 19, 28
10 E&RFTS 42
11 FTS 40, 43
13 FTS 40

14 MU 14
14 FTS 41
15 FTS 39
41 Sqn 156
53 OTU 42
54 OTU 38
60th Rifles 49
61 OTU 42
65 Sqn 155
66 Sqn 37
71 OTU 41
74 Sqn 25,41
II./ZG 76 159
79 Sqn 36
92 Sqn 19, 21, 28, 29, 32, 34, 35, 37, 39, 40,
    43, 45, 47, 49, 52, 65, 117, 118, 152, 155

125 Sqn 38
145 Sqn 41, 42, 155
151 Sqn 42
154 Sqn 41, 42
193 Sqn 43
229th AA Bty, RA 49
231 OCU 39
232 Wing 38
257 Sqn 43
421 Flight 37
504 Sqn 39
601 Sqn 36
615 Sqn 29, 30, 35

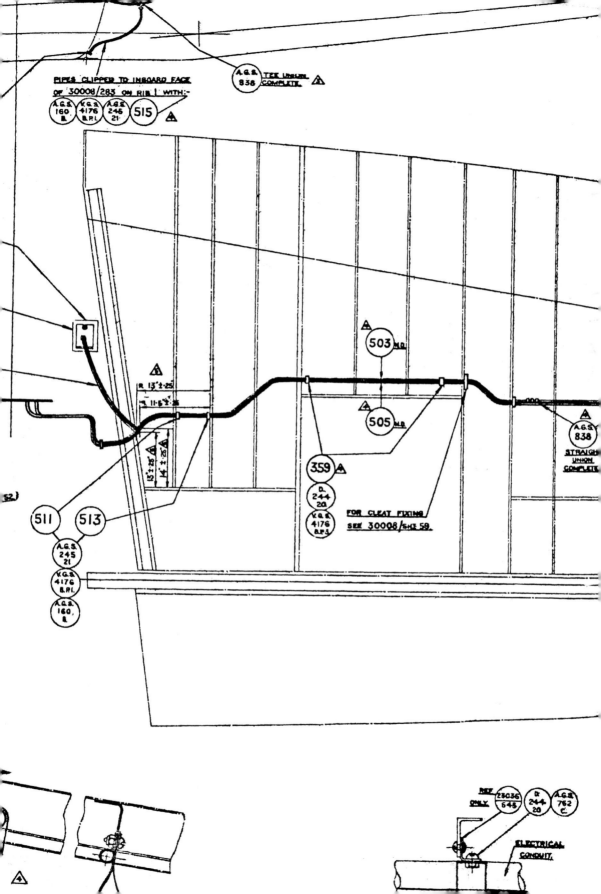